D1264966

# HERBIVOROUS INSECTS

## Host-Seeking Behavior and Mechanisms

# HERBIVOROUS INSECTS

## Host-Seeking Behavior and Mechanisms

EDITED BY

Sami Ahmad
*Department of Entomology and Economic Zoology*
*Cook College, New Jersey Agricultural Experiment Station*
*Rutgers—The State University*
*New Brunswick, New Jersey*



1983

ACADEMIC PRESS
*A Subsidiary of Harcourt Brace Jovanovich, Publishers*
New York London
Paris San Diego San Francisco São Paulo Sydney Tokyo Toronto

ACADEMIC PRESS, INC.
111 Fifth Avenue, New York, New York 10003

*United Kingdom Edition published by*
ACADEMIC PRESS, INC. (LONDON) LTD.
24/28 Oval Road, London NW1 7DX

**Library of Congress Cataloging in Publication Data**

Main entry under title:

Herbivorous insects.

Includes bibliographies and index.
1. Insects--Behavior. 2. Insects--Host plants.
I. Ahmad, Sami.
QL496.H43 1983 595.7052'49 82-20717
ISBN 0-12-045580-3

PRINTED IN THE UNITED STATES OF AMERICA

83 84 85 86 9 8 7 6 5 4 3 2 1

# Contents



## Part I. Neurophysiological Aspects



## Part II. The Diversity of Behavioral Cues

7. Host-Seeking Behavior of the Gypsy Moth: The Influence of Polyphagy and Highly Apparent Host Plants

DAVID R. LANCE



## Part IV. Evolutionary Aspects of Host Selection

8. Selective Factors in the Evolution of Host Choice by Phytophagous Insects

DOUGLAS J. FUTUYMA

# Contributors

Numbers in parentheses indicate the pages on which the authors' contributions begin.

SAMI AHMAD (173), Department of Entomology and Economic Zoology, Cook College, New Jersey Agricultural Experiment Station, Rutgers—The State University, New Brunswick, New Jersey 08903

MAUREEN CARTER (27), Section of Ecology and Systematics, Division of Biological Sciences, Cornell University, Ithaca, New York 14853

V. G. DETHIER (xiii), Zoology Department, Morrill Science Center, University of Massachusetts, Amherst, Massachusetts 01003

PAUL FEENY (27), Section of Ecology and Systematics, Division of Biological Sciences, Cornell University, Ithaca, New York 14853

DOUGLAS J. FUTUYMA (227), Department of Ecology and Evolution, State University of New York at Stony Brook, Stony Brook, New York 11794

FRANK E. HANSON (3), Department of Biological Sciences, University of Maryland—Baltimore County, Catonsville, Maryland 21228

DAVID R. LANCE (201), Department of Entomology, University of Massachusetts, Amherst, Massachusetts 01003, and United States Department of Agriculture, Animal and Plant Health Inspection Service, Otis Methods Development Center, Otis ANG Base, Massachusetts 02542

GERALD N. LANIER (161), Department of Environmental and Forest Biology, College of Environmental Science and Forestry, State University of New York, Syracuse, New York 13210

MICHAEL L. MAY (173), Department of Entomology and Economic Zoology, Cook College, New Jersey Agricultural Experiment Station, Rutgers—The State University, New Brunswick, New Jersey 08903

DANIEL R. PAPAJ (77), Department of Zoology, Duke University, Durham, North Carolina 27706

MARK D. RAUSHER (77), Department of Zoology, Duke University, Durham, North Carolina 27706

LORRAINE ROSENBERRY (27), Section of Ecology and Systematics, Division of Biological Sciences, Cornell University, Ithaca, New York 14853

MAUREEN L. STANTON* (125), Department of Biology, Yale University, New Haven, Connecticut 02511

*Present address: Department of Botany, University of California, Davis, California 95616.

# Preface

With herbivorous insects the sequence of selecting a host is catenary, beginning with recognition of the host plants from a distance, followed by arrival of the insect at the host and subsequent feed and/or oviposition. During this process, several environmental cues interact at the different behavioral levels that eventually result in host acceptance. The material presented in this book addresses itself to the initial aspects of this chain of events, specifically to mechanisms of searching behavior leading ultimately to host location. Our intention is to synthesize current data and ideas on host location research for behaviorists, ecologists, entomologists, evolutionary biologists, and physiologists.

The topics in this volume are divided into four sections: neurophysiology; the diversity of behavioral induction cues; searching mechanisms as affected by insects' breadth of diet; and, finally, an evolutionary analysis of the behavioral and physiological adaptations in insect/host plant relations. Clearly, this volume is not an end point in the development of scientific thought on host location by herbivorous insects, but we hope that it will provide direction toward developing a unifying theme and improving our ability to unravel the complexities of insect/plant interactions.

I am particularly grateful to the staff of Academic Press for their invaluable help with this publication, to Dr. V. G. Dethier (Zoology Department, University of Massachusetts, Amherst) for writing the Introduction, and to the authors for the thoroughness of their contributions. In many cases the reviews are combined with the results of many years of research which are presented for the first time. The authors also made many useful suggestions during the early stages of organizing this volume. They were very cooperative and understanding, and their help made editing this book a pleasant task.

I would also like to thank Mrs. Evelyn Weinmann and Miss C. von Gruchalla for skillfully typing a number of sections of this volume. Thanks also are due to my colleagues Dr. J. H. Lashomb and Dr. M. L. May for many valuable discussions and to Dr. H. T. Streu for his enthusiastic support of my efforts. The editorial work was supported in part by a grant (URF-G-82-370-NB-11) from the Research Council of Rutgers University. This publication (No. C-08130-01-82) was also partially supported by state funds and Hatch Act funds (NJAES-08130) through the New Jersey Agricultural Experiment Station. The authors also were supported by several grants that are acknowledged in their chapters.

Sami Ahmad

# Introduction

Every herbivorous insect is associated with a specific range of host plants. Regardless of the breadth of this range, whether it consists of a single species of plant or encompasses many, the maintenance of stable host/insect relations and the evolution of new ones depends first on the ability of insects to find the plants with which their species is historically associated. These plants must be discovered and identified against a background of many irrelevant species within diverse and often varying quantitative and spatial vegetational contexts.

Having located the host plant in space, the herbivorous insect verifies its identity before ovipositing or feeding. Host selection thus becomes a matter of search and assessment. Each of these processes is effected by an orderly sequence of behavior patterns steered by successive stimuli relevant to the situation. This global view of host-plant selection has required no fundamental revision since it was expounded thirty years ago.

During the intervening period, the subject of insect/plant relations has stimulated the organization of numerous meetings to discuss fundamental issues. The impetus to encourage discussion began with a symposium on insect/plant relations organized by Jan de Wilde in conjunction with the IXth International Congress of Entomology convened in Amsterdam in 1951. This symposium proved to be the genesis of a continuing series of quadrennial Insect Plant Symposia held in Wageningen, The Netherlands in 1957, 1969, and 1982, in Budapest in 1974, and in Slough, England, in 1978. In 1980 a Gordon Conference on Chemical Aspects of Plant–Herbivore Interactions was held in Santa Barbara, California. In addition to the published proceedings of the quadrennial international symposia the following works have appeared: Insect/Plant Relationships, a Symposium of the Royal Entomological Society of London, 1973; Comportement des Insectes et Milieu Trophique, Tours, France, 1976; Biochemical Aspects of Plant and Animal Coevolution (J. B. Harborne, ed.), 1978; and Herbivores (J. A. Rosenthal and D. H. Janzen, eds.), 1979.

A perusal of these volumes will reveal that although the matter of locating host plants in the environment was never neglected, it was subordinated, in terms of research effort expended, to other aspects of host-plant selection. That this relative neglect occurred is hardly surprising considering the enormous difficulties attendant upon observing the behavior of individual flying insects in their natural environment, a difficulty that is amply illustrated in this volume and one that still challenges progress. It is salutary, therefore, that an attempt has been made to focus on the search phase of host selection. This volume is a sampling of current research in that area. While the number of contributions is necessarily restricted, their combined bibliographies provide a comprehensive view of current knowledge.

Of the eight contributions, six deal directly with host-seeking behavior and allied patterns of behavior. The first paper serves as an introduction to the chemical sensory system as it relates to host selection in general. The concluding paper examines evolutionary aspects of the process.

The ability of an insect to find and assess a plant obviously depends on an appropriate sensory system conveying the requisite quantity of information about the environment. The sensory systems most intimately involved are the visual and the chemical. Volumes have been written about these two sensory modalities, but knowledge of how they operate in the field with respect to plants and what the natural adequate stimuli are is still rudimentary.

A paucity of information that becomes particularly restrictive is that relating to differences in visual capacities among different species of insects and, more particularly, how different species perceive their botanical environment. Data describing spectral sensitivity, form perception, and edge and contrast perception, to mention but a few characteristics of visual systems, are scanty. Even less complete are data describing hue, saturation, reflectance, transparency, and other optical properties of leaves.

As Feeny, Rosenberry, and Carter note in their discussion of oviposition by swallowtail butterflies, the weight to be assigned to vision in orientation and recognition varies even among closely related insects. Furthermore, different phases of the coordinated search procedure may depend primarily on vision or primarily on olfaction. Several of the contributors to this volume refer to the lacunae in this area of research; it is regrettable that this subject has not been explored more fully at this time.

The difficulty of relating intrinsic sensory capacities to behavior in the field also appertains to olfaction. Insofar as this sense relates to oviposition, evidence from Feeny's studies of swallowtail butterflies indicates that specific chemicals acting as ''token stimuli'' play a dominant role. On the other hand, as May and Ahmad point out, studies of orienting behavior of Colorado potato beetles indicate that a ratio of nonspecific compounds representing a profile of a plant's essence constitute the perceived stimulus.

Interpretation of behavior toward odors is further complicated by uncertainty regarding the characteristics of odor plumes. Recent studies of the internal structure of odor plumes indicate that a flying insect experiences a far more complex stimulus situation than formerly assumed.

Another aspect of response to odors of botanical origin, which is discussed by Lanier, is the contribution made by members of the species that have already found the host. With bark beetles, in particular, pheromones are active constituents of the total stimulus complex. The searching beetle presumably is required to integrate information from pheromone receptors, receptors responsive to volatile constituents of plants, and visual receptors.

Some perceptions of the intrinsic capacities of the chemical senses of herbivorous insects, the nature of the stimuli, and the potential information carrying capacities of the system are described by Hanson. As Lance has pointed out, however, we still are ignorant of the volume and kind of information about a total plant that are actually gleaned by the chemical senses. For example, not all toxic substances are detected by gustation or olfaction; nor are stimuli mediating deterrence necessarily indicative of a nutritionally unsuitable plant. Furthermore, chemical senses are not able invariably to provide complete assessments of nutritional suitability. Postingestive physiological mechanisms that influence locomotion, and hence dispersal, may operate effectively in plant selection where sensory systems fail.

While the intrinsic capacities of visual and chemosensory systems set the upper limits of perception, the context in which stimuli are presented in nature sets the actual limits. Regardless of whether host location is achieved by visual or chemical cues or by some combination of both, environmental features are limiting factors. As Stanton describes, the discovery of host plants depends very much on circumstances of host distribution in space, the size of stands, the diversity of species within a stand, the identity of nonhost species in mixed stands, and the edge characteristics of a patch. It depends also, as Stanton, Lanier, May, Ahmad, and Lance have pointed out, on whether the searching insect is a specialist or a generalist, adult or immature. It further depends on a nexus of transient variables in any given individual insect, as Papaj and Rausher have outlined.

In perusing these eight presentations one notes the recurrence of familiar fundamental questions, themes, dilemmas, and frustrating lacunae, some of which emerged repeatedly in earlier literature and remain to obstruct our understanding. Designation of the breadth of host acceptability and suitability is one problem. It is clear, as several contributors have emphasized, that patterns of search behavior differ for specialists and generalists. What constitutes a specialist or a generalist is itself unclear because, as pointed out long ago, the terms monophagy, oligophagy, and polyphagy are arbitrarily selected points along a continuous spectrum. In any case, the specialist's task of locating a single species of plant in a mixed floral community requires different tactics than those that are effective for a generalist capable of accepting more than one alternative species. It is in this context that generalizations may fail and that models designed to delineate efficiencies of particular search tactics must take into account breadth of diet.

Another recurring theme relates to variations introduced into behavioral responses by the environment, by the host plant, and by changes within the insect itself. The host plant can, for example, influence such components of orientation as rate of turning, or the insect can modify its behavior as a consequence of experience. Papaj and Rausher discuss these and other modifying conditions. Among internal factors are age, ovarian cycle, hormonal cycles, level of satiation, and competing events such as tendency to migrate or mate.

The emphasis in this volume is on host-finding behavior and mechanisms. If one did not already realize it before reading these articles, it becomes clear that host-finding behavior must be studied in relation to the total behavior pattern of which it is a part. Then the forbidding complexity and diversity of the host/herbivore relationship reveals itself. Can one hope to extract from the mass of data any valid generalizations or unifying theme? Futuyma in the concluding article is not very optimistic. Considering two alternative possibilities, a mechanistic theme or an evolutionary theme, he believes that there is more hope of discovering the latter. He examines cost–benefit models and optimization models.

Whenever one may seek a unifying theme in herbivore/plant relations, whether it be evolutionary or otherwise, more information about mechanisms is sorely needed. Only then will it be possible to assemble knowledge gained from studying separate aspects of total coordinated behavior patterns. The great service that this volume provides is that of focusing for the first time on the initial phases of host/plant relations, that is, on search and location. It thus provides a stimulus for further work in this challenging area. Furthermore, by examining this aspect of behavior in the context of the total insect/plant relationship, it takes its place among those volumes that are indispensable reference works for graduate students and established scholars and for specialists and generalists alike.

V. G. Dethier

# PART I

# Neurophysiological Aspects

# 1

# The Behavioral and Neurophysiological Basis of Food Plant Selection by Lepidopterous Larvae*

FRANK E. HANSON



## I. INTRODUCTION

Host-plant selection by insects in their feeding stages is composed of three distinct behavioral phases: (1) attraction to a potential food plant, (2) arrest or cessation of locomotion, and (3) stimulation (or deterrence) of feeding on that plant. Although visual and mechanical stimuli are utilized to some extent in various phases of the overall feeding behavior, the primary agents controlling

*Much of the research from our laboratory was supported by the U.S. Department of Agriculture, the National Science Foundation, and the Whitehall Foundation.



HERBIVOROUS INSECTS

ISBN 0-12-045580-3

these interactions are the phytochemicals. Thus the critical interface between plant and insect is between the plant chemicals and insect chemoreceptors, the understanding of which will be necessary for an explanation of the basic mechanisms of herbivory.

Accordingly, the ensuing discussion of current knowledge and new directions in the study of behavioral and physiological mechanisms of host selection will follow three approaches: (1) description of the herbivore's feeding behavior, (2) identification of the chemicals in the host plant that elicit feeding behavior, and (3) determination of the physiological responses of insect chemoreceptors to these chemicals.

## II. BEHAVIORAL RESPONSES TO POTENTIAL FOOD PLANTS

### A. Preference Levels for Host and Nonhost Plants

All plants contacted elicit behavior from phytophagous insects: some are rejected, some accepted. Although host plants are in the latter category, they are not all equally preferred; furthermore, some nonhosts are acceptable as well. This raises the question as to where the boundary should be drawn between host and nonhost. Probably the best answer is the ecological one: host plants are those on which the animal completes normal development in nature. The insect's "physiological host range," however, includes some nonhost plants, because many of these elicit feeding and are nutritionally adequate. It follows that screening by the sensory and central nervous systems must be sufficiently broadly tuned to accept a wider group of plants than just those on which the animal is found in nature. Perhaps a continuum of preferences exists, with rejected nonhosts located beyond some hypothetical threshold of acceptability. This threshold is poorly defined and fluctuates with environmental conditions and prior feeding experience.

The factual basis for the previous discussion has been somewhat sketchy; adding flesh to this skeleton requires behavioral assays that permit quantitative comparisons of the feeding preferences for all plants tested. Early efforts in quantitative measurements of feeding tended to be subjective, with feeding estimates scaled from 0 to 5 or with pluses and minuses. An improvement was introduced by the disk test (Fig. 1), which allows quantitative comparisons of preference for different plant species (using leaf disks) or chemicals (using filter-paper disks) (Stadler and Hanson, 1978). By keeping one plant or a control chemical common, the preferences of many plant species or chemicals can be compared quantitatively.

Evidence that all host plants are not equally preferred has been obtained using

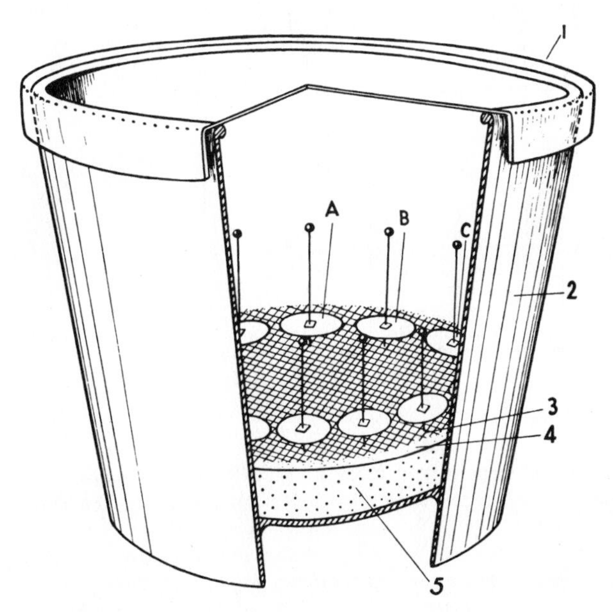


**Fig. 1.** The disk test. Disks cut with a cork borer from leaves of species A, B, and C are mounted on pins and held by small acetate squares 1 cm above the wax substrate. 1, Cover; 2, plastic or paper cup; 3, wire screen; 4, moist filter paper; 5, paraffin wax layer. The caterpillar is placed in the center; feeding is scored as area of disk eaten. (From Jermy *et al.*, 1968; used with permission.)

this method. For example, larvae of the tobacco hornworm (*Manduca sexta*), reared on tomato leaves, were given a choice between leaf disks of tomato and another plant species. The results were normalized to the consumption of tomato. The relative feeding preferences for nine different host solanaceous plants are shown in Fig. 2. All of these host plants are acceptable, although considerable differences are evident between the most and the least preferred of the solanaceous plants.

The discrimination between host and nonhost can also be seen in Fig. 2. The cruciferous plants rape and radish and the legume cowpea have feeding scores that are comparable or lower than the least preferred of the tested solanaceous plants. Nevertheless, larvae can be reared successfully on these nonhosts in the laboratory, showing that they can support larval development. Less acceptable nonhosts also can be tested and compared quantitatively (Fig. 3).

Preference hierarchies for the tobacco hornworm were previously reported for a wide variety of solanaceous plants by Yamamoto and Fraenkel (1960c). Although the methods used were different, similar conclusions were drawn.

In summary, there is evidence that the tobacco hornworm discriminates among host plants as well as between host and nonhost. As suggested by the theoretical

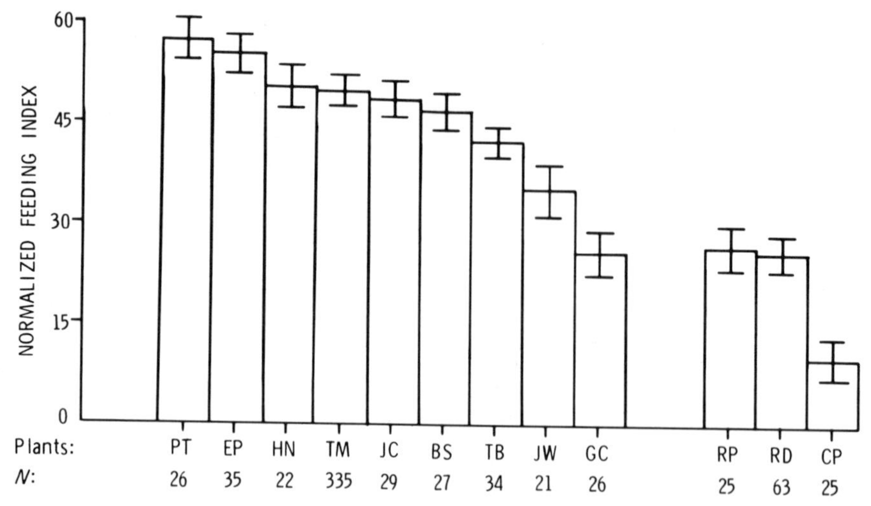


**Fig. 2.** Relative feeding preferences of *Manduca sexta* for nine solanaceous (host) plants and three nonsolanaceous (nonhost) plants. Solanaceous plants: PT, potato, *Solanum tuberosum;* EP, eggplant, *Solanum melongena;* HN, horse nettle, *Solanum carolinense;* TM, tomato, *Lycopersicon esculentum;* JC, Jerusalem cherry, *Solanum pseudocapsicum;* BS, bittersweet, *Solanum dulcamara;* TB, tobacco, *Nicotiana tabacum;* JW, Jimsonweed, *Datura stramonium;* GC, ground-cherry, *Physalis* spp. Nonsolanaceous plants: RP, rape, *Brassica napus;* RD, radish, *Raphanus sativus;* CP, cowpea, *Vigna sinensis*. *N*, Number of animals tested on this plant. Columns represent mean feeding scores for *N* animals, normalized by setting the mean tomato-feeding score at 50. Bars at top of columns are ±SEM. All animals reared on tomato and tested in two-choice disk tests, each plant versus tomato. (Data from G. deBoer and F. E. Hanson, unpublished, 1982.)

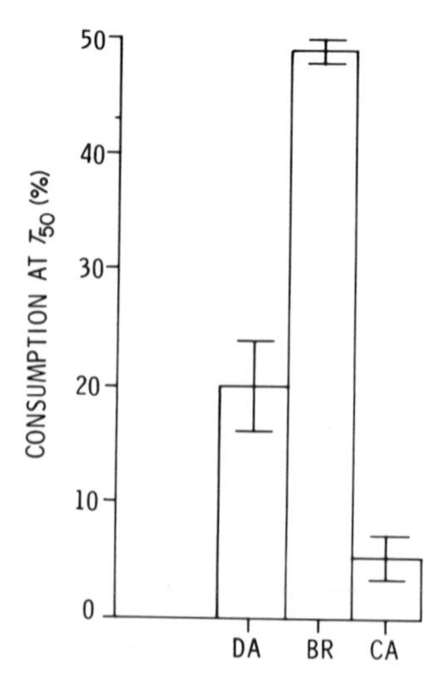


**Fig. 3.** Feeding preferences among three nonhosts. Ordinate: percentage of leaf disk area eaten by $T_{50}$, the time at which consumption of any plant reached 50%. DA, Dandelion, *Taraxacum officinale;* BR, cauliflower, *Brassica oleracea botrytis;* CA, canna, *Canna indica*. For comparison with Fig. 2, the *Brassica* columns in each figure are comparable. Animals were reared on wheat germ medium. $N = 25$. (After Jermy *et al.*, 1968.)

treatment of oligophagy by Jermy (1961), a graded continuum of acceptability appears to extend across host and nonhost plants, rather than there being a sharp demarcation between highly acceptable hosts and unacceptable nonhosts.

## B. Behavioral Plasticity in Feeding Preference

Prior feeding on a particular food plant may influence an insect to select that rearing plant preferentially. Johansson (1951) found that fifth-instar larvae of *Pieris brassicae,* the European cabbage butterfly, preferred the plants from which they had been collected in the field. Laboratory studies confirmed the existence of induced preferences in the tobacco hornworm and the corn earworm, *Heliothis zea* (Yamamoto and Fraenkel, 1960b; Jermy *et al.*, 1968). Eight other species of lepidoptera have subsequently been shown to be subject to induced preferences, which suggests the phenomenon may be widely distributed (Ting, 1970; Hanson, 1976; Greenblatt *et al.*, 1978). Induction is not, however, completely universal; lack of induction has been reported in *Pieris rapae* and *Pieris napi* (Chew, 1980).

Comparisons of the strength of induction with the degree of polyphagy show a positive correlation (Hanson, 1976). For example, the larva of the relatively polyphagous promethia moth, *Callosamia promethia,* shows a very strong induction of preference (Fig. 4). This figure graphically depicts that strong preferences can be induced in opposite directions by different feeding experience: for example, animals reared on cherry will almost completely ignore sassafras, and vice versa. In contrast, the oligophagous *M. sexta* shows a smaller effect of prior

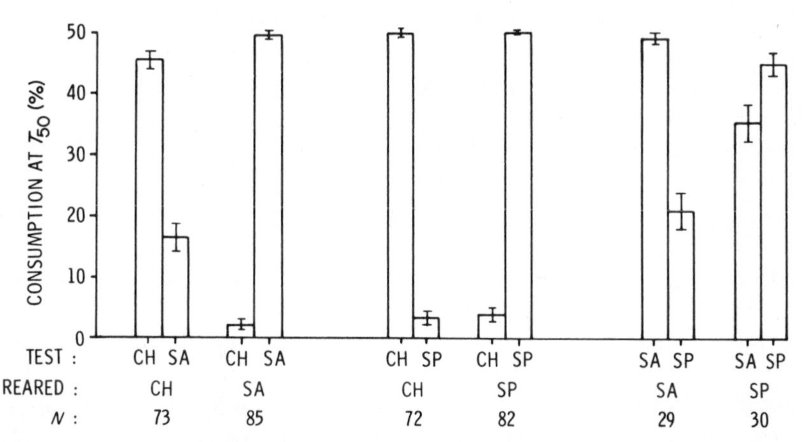


**Fig. 4.** Induction of preference in *Callosamia promethia*. Rosaceae–Lauraceae paradigms show much stronger induction than Lauraceae–Lauraceae. CH, Cherry, *Prunus serotina* (Rosaceae); SA, sassafras, *Sassafras albidum* (Lauraceae); SP, spicebush, *Lindera benzoin* (Lauraceae) ($p << .001$). (Data from F. E. Hanson, unpublished.)

experience on feeding preference, and the more narrowly oligophagous *Pieris* apparently show none.

Although the correlation between polyphagy and induction is clearly present, recent data have led to an expanded viewpoint. For example, some insects are induced strongly with some plant pairs, but less so or not at all with others. Taking these data into account, Wasserman (1982) and G. deBoer and F. E. Hanson (unpublished data) correlate the strength of induction with the taxonomic relatedness of the inducing plants rather than the degree of polyphagy of the insect that is induced. Thus, as shown in Fig. 4, a strong preference was induced in promethia by both cherry and sassafras, which are in different families (Rosaceae and Lauraceae, respectively) and therefore may be quite different chemically. The same is true with cherry and spicebush. But the sassafras and spicebush pair, which are both Lauraceae, show a much weaker induction. The lack of induction by *P. rapae* and *P. napi* on host food (Chew, 1980) may simply reflect that the experimental foods tested were all closely related from one family (Cruciferae). Chemical similarity between inducing plant pairs may result in similarly induced animals, or the plants may be more difficult to discriminate during the choice test. Lack of induction does not necessarily mean lack of discrimination, as cases have been found where the animals definitely discriminated between two plants that did not cause an induction of preference (G. deBoer and F. E. Hanson, unpublished data).

Techniques other than the disk test also manifest induced preferences. Saxena and Schoonhoven (1982) measured total feeding by animals placed on a single food: animals reared on that food ate more of it than those reared on a different food. Thus the induction of preference is not likely to be an artifact of procedure. The chemical and physiological basis of the induced preferences have only begun to be elucidated. Leaves per se are not required, because preferences can be induced by rearing on artificial media: just as in induction with leaves, the animals prefer the media on which they have been reared (Stadler and Hanson, 1978; Ave, 1981). One experiment showed induction for linolenic acid, suggesting that low molecular weight lipids may play a role in induction.

The physiological basis of induced preference is not known, but suggested mechanisms include changes in sensory or central nervous systems. The chemosensory organs are involved because sensory ablation eliminated the preference or its manifestation (Hanson and Dethier, 1973). Another indication that its basis may be in the peripheral sensory systems comes from the data of Schoonhoven (1969) and Stadler and Hanson (1976): electrophysiologically recorded responses of maxillary styloconica to plant-derived stimuli are not the same in tobacco hornworms reared on different plants. These studies implicate the sensory systems but do not exclude additional changes in the CNS. Other explanations have also been proposed, such as locomotion arrest upon encountering a type of food previously ingested, or aversive digestive effects that curtail feeding on plants on which the animal has had no previous feeding experience. These hypotheses remain to be tested.

The importance of induced preferences to the animal is not well understood. Yamamoto (1974) suggested that an advantageous oligophagy is induced in a genetically polyphagous animal. Although this is not a permanent change, because it can be replaced by forced feeding on another plant, it must nevertheless exert a statistically important effect when a feeding choice is required. In a natural habitat where leaves and branches of different plants intermingle, for example, larvae may contact several species during normal feeding. Induction of preference may then tend to restrict feeding to one host species. Such restriction may be unimportant for some species (as discussed by Jermy *et al.,* 1968) but can be crucial to survival in others. For example, if promethia larvae reared on sassafras are provided cherry as the sole food, a high mortality will result (55–95%, depending on season). Presumably, the cherry leaves contain a toxic agent (the cyanogen amygdalin is a possible candidate) that cannot be detoxified by animals reared on sassafras. Thus it would be highly advantageous to these animals to ignore cherry, which they do (Fig. 4). In contrast, animals reared on cherry apparently maintain the capability to detoxify cherry; they prefer it (Fig. 4) and develop normally on it. In this case at least, and probably in many others as well, the induction of preference is clearly adaptive.

Awareness of the induction of preference is of paramount importance to the experimenter measuring feeding behavior (Wiseman and McMillan, 1980). Because induced food preferences may mask hereditary preferences, methodological difficulties will be encountered when preference tests are used to determine natural host ranges if the rearing plant is not accounted for.

Similarly, insects reared on different foods will have different responses to test stimuli in experiments investigating the chemical basis of food choice (Stadler and Hanson, 1978). This adds another dimension (that of prior feeding experience) to the problem of experimentally determining whether plant extracts and phytochemicals are stimulants or deterrents for the insect. Such an example is shown in Fig. 5. For animals reared on wheat germ medium (Fig. 5A), feeding responses to the hexane extract of wheat germ medium were comparable to the solvent control, suggesting that this extract contains neither stimulants nor deterrents and is therefore "neutral." However, to animals raised on a different food (e.g., tomato, Fig. 5B), this same extract is deterrent compared with controls. Thus erroneous conclusions about the deterrency of the extract would have been drawn if only diet-reared animals (Fig. 5A) had been tested.

A similar experimental use of the induction of preference is to determine whether the animal can differentiate between two stimuli appearing equally acceptable. For example, tomato and solanum are both host plants that are equally accepted by animals reared on artificial diets (Jermy *et al.,* 1968). Only by raising them on the two host plants and comparing feeding responses can one

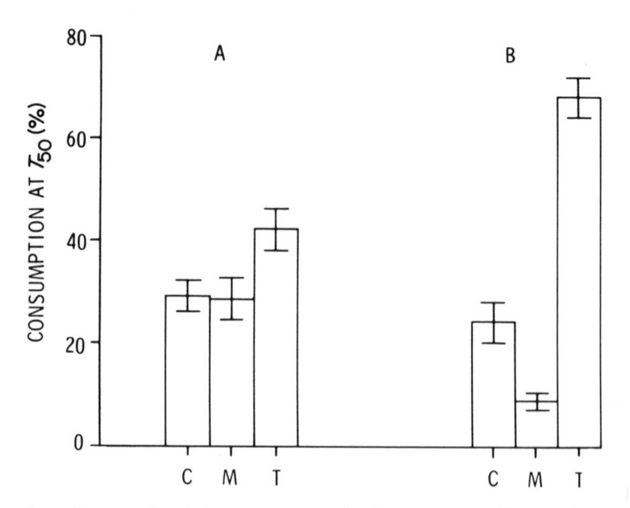


**Fig. 5.** Induced preferences for hexane extracts of wheat germ medium and tomato leaves in *Manduca sexta*. Larvae were reared on (A) wheat germ medium (B) or tomato leaves and tested using glass fiber filter-paper disks impregnated with hexane extracts of wheat germ medium (M), freeze-dried tomato leaves (T), or the solvent alone as the control (C) ($N = 20$; $p < .001$). (Adapted from Stadler and Hanson, 1978.)

demonstrate that the caterpillars possess the ability to differentiate between the two.

The induction of preference can also be used to assess whether leaf fractions contain host-plant cues unique to a particular plant species or whether they simply contain general stimulants or deterrents. This is because induced preferences apparently are not manifest for general stimulants, such as sucrose and inositol (Stadler and Hanson, 1978), but are exhibited for plant-specific stimuli (e.g., see Fig. 7).

Thus, the very qualities of the induction of preference that signal caution for some investigations also permit it to be used as an important experimental tool.

### C. The Role of Olfaction in Larval Food Selection

Olfactory attraction to food plants clearly is more important for highly mobile adults than for the relatively sluggish larvae. Nevertheless, food choice is of such importance to larvae that all possible detection mechanisms are likely to be employed. The vapor-phase chemical mixture near a leaf is probably specific to each plant species and would likely be an important factor in the animal's discriminant analysis.

The ability of larvae to orient toward host plant leaves can be shown experimentally by a simple two-choice test using a large-mesh wire grid. Two leaf disks (or filter-paper disks, for chemical studies) of uniform shape and size are placed vertically 1 cm to each side of the center wire at a grid crossing point. A larva placed on the center wire moves along to the intersection, at which point it may continue forward (no choice) or turn onto the transverse wire toward one of the choices. A comparison of the number of larvae turning toward one or the other leaf disk represents a measure of the differences in attractivity and can be analyzed with binomial statistics. Using this technique, Saxena *et al.* (1976) showed unequivocally that larvae were able to select their host plants from 1 cm away. The chemical basis of this olfactory-mediated attraction can be examined with this assay using filter-paper or agar disks laced with chemicals (Saxena and Schoonhoven, 1978) or extracts of leaves (Section III).

Despite these obvious demonstrations that distance olfaction can serve to orient and attract caterpillars to a potential food source, it does not seem to be as important in actual food selection as is gustation. Ablation of olfactory organs diminishes the animal's ability to distinguish among host plants (Hanson and Dethier, 1973), but has little effect on host–nonhost discrimination. Ablation of the gustatory organs (styloconica and labial receptors), however, permits feeding on nonhosts, even those that are strongly rejected by the normal animal, thereby greatly expanding the ''physiological host range'' (Dethier, 1953; Waldbauer and Fraenkel, 1961; Waldbauer, 1962, 1964; deBoer *et al.*, 1977).

Another source of ''olfactory input'' may be from the classical gustatory receptors. For example, the lateral maxillary styloconica of the tobacco hornworm have been shown electrophysiologically to respond to cut edges of a leaf as far as 0.5 mm away (Stadler and Hanson, 1975). Perhaps this mechanism accounts for the discrimination that occurs upon initial encounter with the leaf. Careful observation of videotaped encounters indicates that feeding decisions are often made without ''test bites,'' and inspection of the rejected leaf often fails to disclose damage to the leaf.

In conclusion, the olfactory component of larval food-plant selection is an important one. Demonstrated roles of olfaction include mediating the broadley tuned, initial attraction toward a potential food source and facilitating fine distinctions such as distinguishing among acceptable food plants. Additional roles for olfaction, such as driving or activating feeding behavior as shown in *Schistocerca* (Mordue-Luntz, 1979), should be investigated further.

Although this discussion has focused only on lepidopterous larvae, it should be mentioned that in many adult insects olfaction is probably the most important medium of discrimination. For example, adult Japanese beetles rendered anosmic by antennectomy are inferior to normals in both feeding intensity and host location (Ahmad, 1982). Also, oviposition site selection by some butterflies is strongly dependent on olfaction (Calvert and Hanson, in press).

## III. CHEMICAL BASIS OF HOST-PLANT SELECTION

### A. The Search for the Chemicals by Which Larvae Identify Host Plants

The investigation of the basis of larval selection of host plants has attempted to isolate or identify the stimulus that uniquely identifies a host plant. This leads to the question whether the stimulus is a single chemical or a complex set of chemical cues.

A single plant chemical added to an artificial diet can cause an increase in feeding. For example, David and Gardiner (1966) and Ma (1972) have shown that sinigrin and related mustard oil glycosides are effective feeding stimulants for *P. brassicae*. This is an oligophagous insect specific to the Cruciferae, a family of plants unique in having a high concentration of sinigrin and other mustard oil glycosides. Similarly, the hypericin beetle, *Chrysolina brunsvicensis,* which lives on the hypericin plant, is stimulated to feed by a single food-plant–specific compound, hypericin (Reese, 1969). These results provided support for the concept that a single ''secondary'' (nonnutritive) plant chemical may provide the host-specific cues (Fraenkel, 1959, 1969).

Investigations have since attempted to find specific feeding chemicals for other insect species. The tobacco hornworm, for example, feeds on solanaceous plants. Several plant alkaloids (e.g., tomatine, atropine, solanine, and solanocapsin) are specific to this family and thus are likely candidates to be the host-specific chemical stimulant. So far, however, none have been found to stimulate feeding in the disk test (G. deBoer, D. Sorrells and F. E. Hanson, unpublished), irrespective of the rearing plant. Other plant chemicals present in quantity in host leaves also were assayed with negative results: chlorogenic acid, solanosol, 2-tridecanone, and GABA. Thus, although there has been much speculation that specific plant alkaloids or other secondary plant compounds may permit *Manduca* to identify its host plant, the compounds tested so far—those that are present in high concentrations in the Solanaceae—do not seem to play such a role.

The ''primary'' (nutritive) plant compounds also have been tested for feeding stimulation or deterrence. Cholesterol, several amino acids, ascorbic acid, and other assorted nutritive compounds were found to be neutral or slightly deterrent (Stadler and Hanson, 1978). Sucrose (Fig. 6) and inositol are only mildly stimulating. No synergism of sucrose with alkaloids or amino acids was found. None of these compounds is sufficiently stimulatory to account for the observed feeding on host leaves. Furthermore, those causing stimulation (sucrose and inositol) are found in nearly all plants and stimulate feeding in most insects; accordingly, these compounds are not likely to play a large role in host-plant identification, as suggested earlier by Fraenkel (1969).

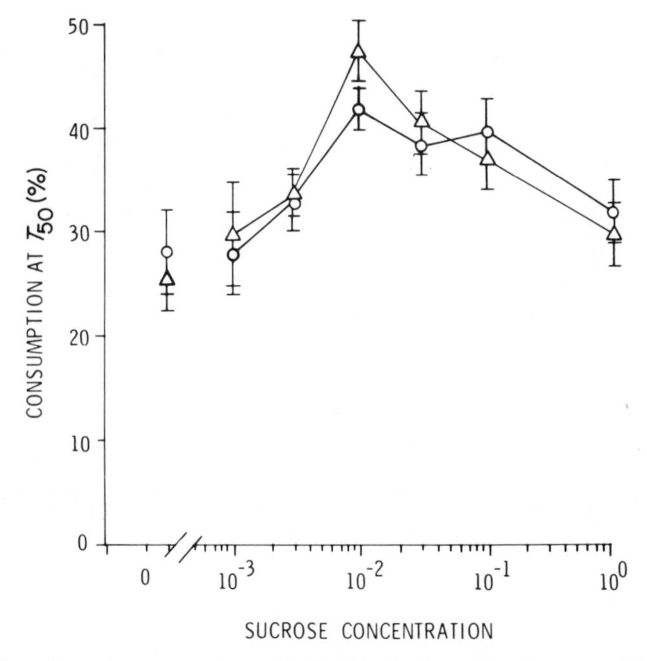


**Fig. 6.** Effect of sucrose concentration on feeding by *Manduca sexta*. Summary of three series of three-choice tests, each using two sucrose concentrations and a water control. Water controls averaged at left. Tested animals were reared on tomato (circles) or Jerusalem cherry (triangles); neither group had significantly different preferences at any single concentration (*t* test). $N = 153$. (After Stadler and Hanson, 1978.)

The only remaining possibility is that low levels of secondary plant chemicals are responsible for providing cues for food-plant discrimination. The number of such chemicals known to comprise leaves of solanaceous plants is myriad. Assaying single chemicals individually in the disk test would be like the proverbial search for the needle in the haystack; therefore another approach to finding stimulants and deterrents in leaves seems more practical.

### B. Chemical Fractionation of Host-Plant Leaves

The second approach to finding chemical stimulants begins with the natural stimulants, namely leaves, which are chemically extracted and fractionated to isolate the active principle. Each fraction is assayed behaviorally for activity as a feeding stimulant or deterrent, and is also used as a stimulus in electrophysiological studies of the sensory organs (Section IV).

Previous work using the chemical fractionation approach showed that extracts of tomato leaves could elicit biting by the tobacco hornworm (Yamamoto, 1957;

Yamamoto and Fraenkel, 1960a). The extracts or fractions were spotted around the perimeter of a single filter-paper circle in a petri dish. A fifth-instar caterpillar was placed in the petri dish. Damage to the filter paper from biting provided a measure of the efficacy of the stimulant. Activity was found in the methanol, ethanol, and water extracts of host plants; the nonpolar extracts, as well as those containing alkaloids and flavonoids, were reported to be nonstimulatory. Because only positive feeding stimulation can be detected with this method, no deterrents could have been found.

The polar extracts were further purified by Howard (1977), using a bioassay that measures the probability of a first-instar larva being associated with the test fraction. This method probably assays attractants and arrestants more than feeding stimulants. Although the active compound was not identified, several steps of purification concentrated the active fraction from the original water extract.

Nonpolar extracts also stimulate feeding (Stadler and Hanson, 1978). In fact, choices elicited by the nonpolar extracts are similar to those evoked by the leaves from which the extracts were obtained. This suggests that stimulants are present

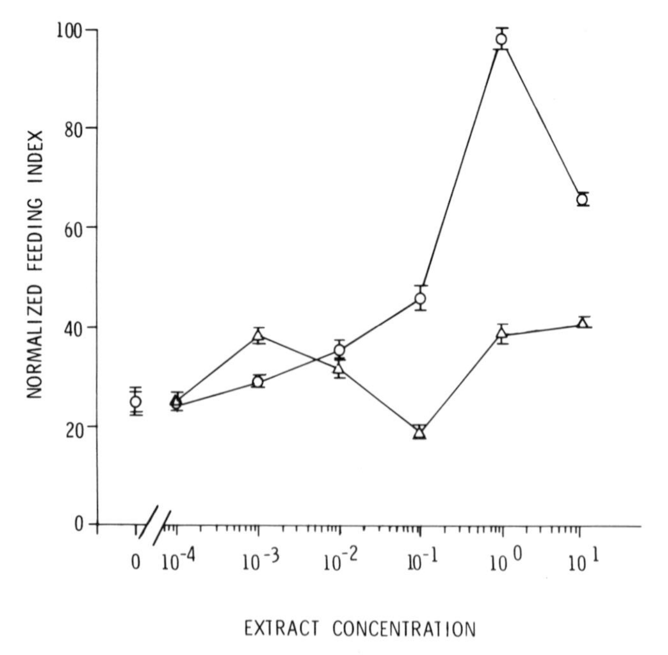


**Fig. 7.** Stimulation of feeding of *Manduca sexta* by nonpolar extracts of tomato leaves. Summary of six series of two-choice tests comparing water versus one concentration of the petether extract. Points are feeding scores normalized to the average scores on the water controls. Extract concentrations are relative to that ($10^0$) which would be present in the portion of a leaf the size of the filter-paper disk used in the tests. Circles, tomato-reared hornworms; triangles, Jerusalem cherry–reared hornworms. $N$ = 315. (Data from deBoer and Hanson, 1982.)

in this fraction that are host specific. The dose–response curve of the nonpolar extracts (Fig. 7, upper curve) shows a maximal feeding stimulation at the concentration normally found in leaves ($10^0$), with decreasing effectiveness at higher and lower concentrations (deBoer and Hanson, 1982). The threshold response occurs at about a 1000-fold dilution. The magnitude of the feeding (about 5 times that of the water control) suggests the presence of a stimulant about 2.5 times as effective as sucrose (Fig. 6).

In contrast to the large-amplitude unimodal dose–response curve obtained with tomato-reared hornworms, a shallow, bimodal response to the same extract was obtained with animals reared on Jerusalem cherry leaves (Fig. 7, lower curve). The latter group of animals was actually deterred (in comparison to responses to water) at a relative concentration ($10^{-1}$) that was stimulating to the former group. This response implies the presence of a feeding deterrent in the extract to which the Jerusalem cherry-reared animals responded, but not the tomato-reared group. The difference in preferences of the two sets of larvae (differences between the two curves of Fig. 7) presents a clear demonstration of induced preference for some component of the tomato extract. This is another example of the importance of considering feeding history when attempting to interpret results of feeding-preference tests (Section II,B). Perhaps further experimentation along this line will elucidate the chemical basis of the induction of preference.

Further fractionation of the petether extract of tomato leaves was accomplished with a silica gel column and successive elution with hexanes, chloroform, acetone, and methanol (deBoer and Hanson, 1982). (Hexanes elute hydrocarbons, chloroform elutes the rest of the neutral lipids, acetone the glycolipids, and methanol the phospholipids plus any remaining polar compounds.) The bioassay on the disk test using tomato-reared larvae showed that the chloroform eluate contained the feeding stimulant, hexanes contained a deterrent, and the acetone and methanol fractions were neutral. Thus we tentatively conclude that the stimulants are neutral lipids. Further fractionation of the active eluates will hopefully lead to purification, isolation, and, eventually, identification of the compounds responsible for stimulating and deterring feeding.

Following the preceding demonstration of an active feeding stimulant in whole leaves, the question arose as to where in the leaf this factor was located. Given the nonpolar nature of the stimulant, the surface wax was thought to be the most likely location. To test this, extracts of tomato leaf surfaces were obtained by dipping whole leaves into chloroform or hexanes for 30 sec to extract the surface chemicals (Hamilton and Hamilton, 1972). The material obtained proved just as active in the disk test as that extracted from whole leaves, suggesting that the feeding stimulant is in the surface wax layer. The "dewaxed" leaves themselves lost some but not all of their stimulatory activity, suggesting that active compounds are also inside the leaves or that the extraction was incomplete.

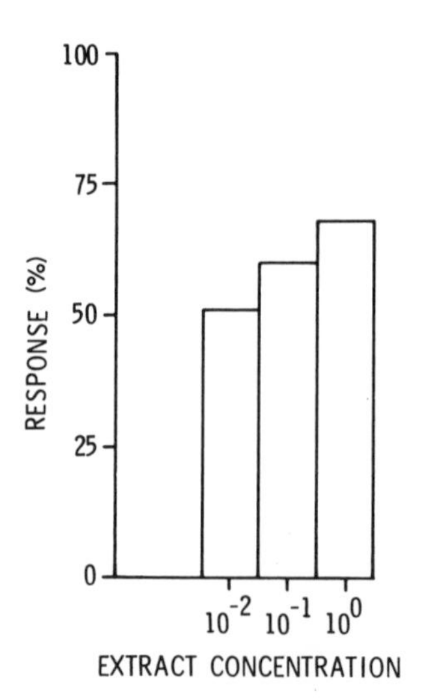


**Fig. 8.** Olfactory attraction of *Manduca sexta* to nonpolar extract of tomato leaves. Choices were glass fiber filter-paper disks laced with either a petether extract of freeze-dried tomato leaves or the solvent alone. Solvents were allowed to evaporate and disks were wetted with 0.1 ml $H_2O$ for the test. See text for explanation. (Data from D. Klein and F. E. Hanson, unpublished.)

The volatile component of the petether extract of tomato leaves also was bioassayed in this laboratory for olfactory attraction. Using the two-choice attraction technique (Section II,C), a stimulatory component was demonstrated (Fig. 8). It is perhaps remarkable that any vapor-phase activity remains after twice drying down the original petether extract before testing; it implies that the olfactory system is very sensitive to food-plant odors and reinforces the contention that olfaction may be quite important in the overall feeding picture.

## IV. RESPONSES OF THE CHEMOSENSORY CELLS TO PLANTS AND PLANT CHEMICALS

The known chemoreceptors on lepidopterous larvae are the paired antennae (olfactory), paired maxillary palpi (mixed olfactory and gustatory), two pairs of maxillary styloconica (primarily gustatory, but also olfactory for short distances), and two gustatory receptors on the labrum. Data discussed earlier (Section II,C) suggest that the styloconica may be the most important discriminatory sense organs and therefore the most logical candidates for investigating chemosensory physiology and stimulus coding.

Histological studies have shown that each of the two maxillary styloconica has four sensory cells. The cell bodies lie beneath the cuticle in the galea with the dendrites projecting through the length of the hollow cuticular peg, ending just below the pore at the tip. It is through this 0.5-μm-diameter pore that the animal interacts with its chemical environment; the tips of the chemoreceptive dendrites lie just beneath, within a few milliseconds of diffusion time from the external chemical world.

Experimental monitoring of the receptor activity is made possible by the diffusion of electrolytes through this pore to provide an electrical connection between the dendrite and the electrode. Accordingly, stimuli containing electrolytes play the dual role of stimulating the sensory dendrite and conducting the electrical currents to the active surface of the recording electrode. At this interface, a change in the concentration of ions in solution leads to a flow of electrons in the solid-state circuitry of the electrode via the electrochemical reaction:

$$Cl^- + Ag \rightleftarrows AgCl + e^-$$

Thus a surge of $Cl^-$ ions in solution arriving as a wavefront at the electrode results in a flow of electrons in the silver–silver chloride surface. Because neural currents are produced and conducted by ionic fluxes, this ionic–metallic junction

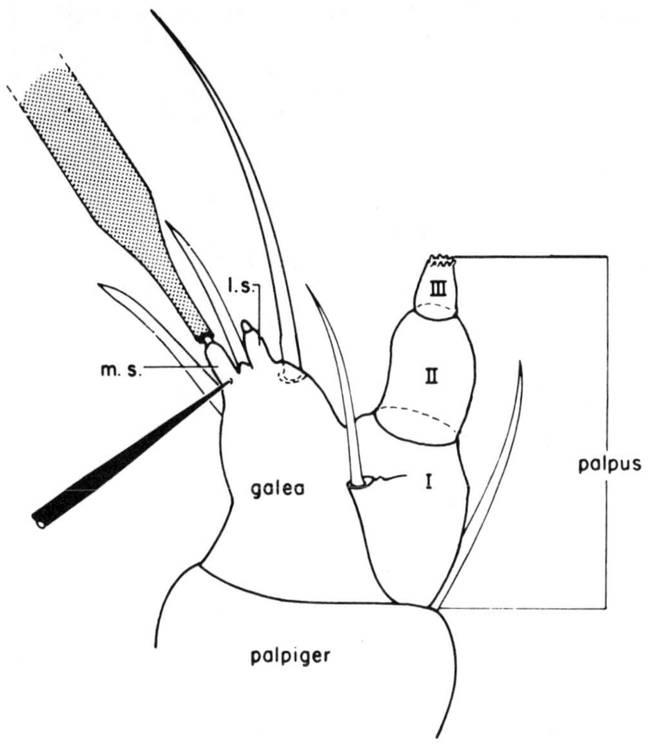


**Fig. 9.** Setup for gustatory recording from maxillary styloconica of *Manduca sexta*. Tip-recording electrode (fluid filled) shown placed over the end of the medial styloconicum; sidewall recording depicted by an electrode inserted near the base of the medial styloconicum; m.s., medial styloconicum; l.s., lateral styloconicum. (After Hanson, 1970.)

provides the means for monitoring activity of the nervous system via electronic instruments. Necessary equipment includes a high-impedance preamplifier, filters, amplifier, and oscilloscope; an FM tape recorder and a computer are useful options.

The physical delivery of the stimulus to the gustatory organs requires a micropipet full of the stimulating fluid, or cut edge of leaf or filter paper, guided by an appropriate micromanipulator. Alternatively, an electrolytically sharpened tungsten wire may be inserted through the cuticle to monitor the activity of the receptor cells chronically, with the stimulus being applied at the tip as before. A diagrammatic view of the recording situation is shown in Fig. 9.

Electrophysiological recordings from the styloconica show considerable receptor cell activity when complex stimuli are applied (Dethier and Kuch, 1971; Dethier, 1973). For example, leaf juice elicited the response shown in the upper tracing of Fig. 10; at least three spike shapes are discernible, suggesting that as many cells are active. If a different plant species is applied to the same styloconicum, a different response is elicited.

Given that plants elicit specific responses from chemoreceptors, the next step

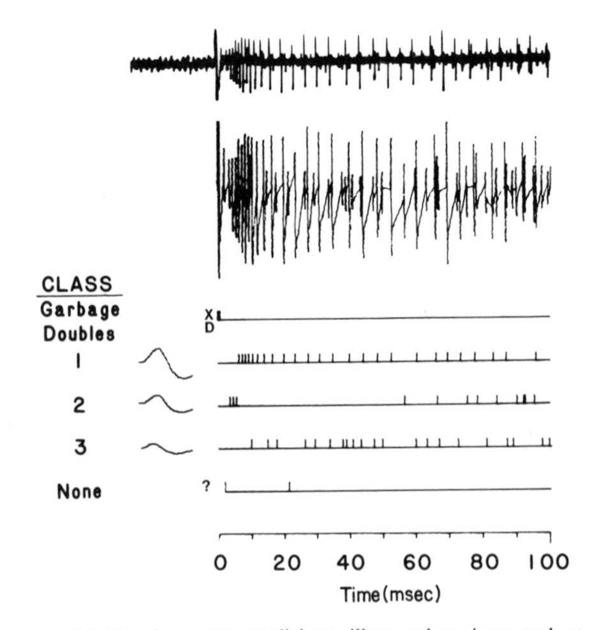


**Fig. 10.** Response of a *Manduca sexta* medial maxillary styloconicum to host-plant leaf juice (tomato). Top trace, filmed analog response, tip-recording technique; next lower trace, computer-reconstructed digitized record demonstrating the fidelity of analog–digital conversion; middle traces, tic marks indicating separation of spikes into the classes denoted by spike templates or symbols at left: X, garbage, D, double, ?, unclassified. Bottom trace, time scale. See text and Fig. 11 for further details. (Data from F. E. Hanson and C. Cearley, unpublished.)

in this study is to determine the sensory code by which the animal's CNS discriminates among plants. The problem is a difficult one, largely because of the variability in recorded responses within and among animals. Consequently, many recordings must be averaged to obtain a representative sample, resulting in a large accumulation of data to be analyzed. Such analysis traditionally requires filming the analog trace and measuring each action potential by hand, a process too laborious to be attempted for more than a few plants. Techniques have now been developed that use the laboratory computer to sort and analyze rapidly large numbers of insect sensory action potentials. For example, the response illustrated in Fig. 10 has been entered into the computer using analog–digital conversion of the action potentials followed by sorting with a template-matching technique (Gerstein and Clark, 1964) that classifies spikes according to their shapes (Hanson *et al.*, in press). The representations of the computer's digital image of the data are plotted just below the filmed analog trace in Fig. 10, with the classification decisions indicated directly beneath each spike. The class and time of occurrence for each spike are then stored for later recall in subsequent analysis and plotting programs. For example, Fig. 11 shows the computer-generated response curves of each class of spikes averaged from nine records (one of which is illustrated in Fig. 10) for each of three plants. The differences in each cell's response to each plant are obvious. These differences appear to be sufficient for the animal to discriminate among these three plants on the basis of input from the styloconicum alone. Statistical analyses (such as used by Dethier and Crnjar, 1982) would further document the differences. When more such data are correlated with behavioral responses of the same animal to the same substances,

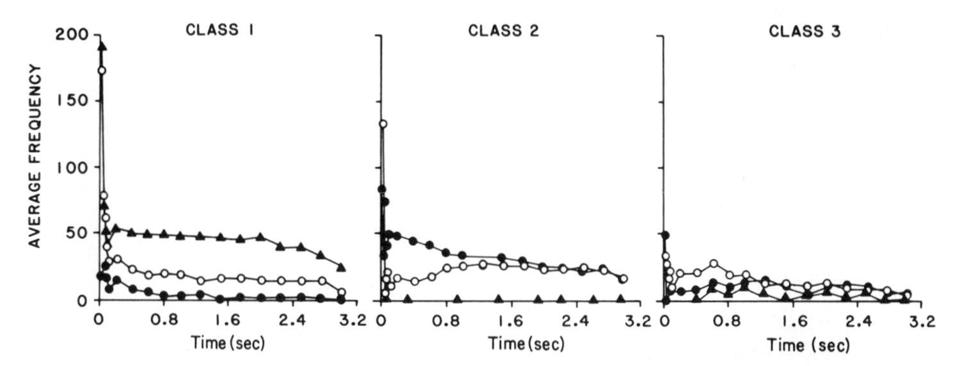


**Fig. 11.** Representation of averaged instantaneous frequencies of individual classes of action potentials. Data from nine trials were classified and averaged by the time-partition histogram method, then converted to instantaneous frequencies. Stimuli were the equally acceptable host plants, tomato (open circles) and tobacco (closed circles), and the unacceptable nonhost *Canna indica* (triangles). Note different rates of activity of each spike class (cell type) from each plant. (Data are from F. E. Hanson and C. Cearley, unpublished.)

perhaps the sensory coding responsible for discrimination among potential food plants may be deciphered.

With these techniques to separate and identify each cell's activity, the response spectrum of each cell within each sensillum ultimately will be characterized. A beginning has been made: Schoonhoven (1973) tentatively concluded that of the eight cells in the two sensilla styloconica, two are sensitive to inositol, two respond to salts and acids, one to sucrose and glucose, one to alkaloids, and one to salicin. To what other compounds these cells are sensitive and the characteristics of their responsiveness remain to be determined.

Once the sensory code for relevant stimuli becomes better understood, perhaps the combination of electrophysiological recording and computer-based spike sorting and analysis can be used to facilitate screening of compounds as possible stimulants or deterrents, just as the use of the electroantennagram has facilitated screening of compounds for pheromonal activity. This may eventually have applied uses for discovering feeding stimulants or deterrents for protection of plants or plant products. These results could find application directly, such as for baits, or indirectly by providing information for plant breeders. The latter will be particularly important as new molecular techniques become available for plant genome manipulation. Regardless of any potential economic benefit, such a successful analysis of sensory coding will be a significant milestone toward complete understanding of the physiological basis of feeding behavior.

## V. SUMMARY

Plant-feeding behavior by lepidopterous larvae is composed of several components: attraction, arrest, and stimulation of feeding. Methods for testing individual components have been developed, such as the two-choice olfaction test and the disk test. Experiments have shown that preference hierarchies exist among host plants, and hosts are generally preferred to nonhosts.

Preference hierarchies thus determined, however, are influenced somewhat by the plant on which the test insects were reared. The primary change is that increased preferences are shown for the rearing plant. This is an example of an induced preference, and demonstrates that, without knowledge of an animal's feeding history, descriptions of food preferences are suspect. The data suggest that their food-choice plasticity can be adaptive by keeping the animal on the same food plant.

One of the long-term goals of understanding insect herbivory is to determine which phytochemicals control feeding by acting as stimulants or deterrents. For some insects, certain host-plant-specific secondary plant chemicals stimulate feeding, suggesting that these could be the basis for host plant specificity for these insects. Attempts to determine such host-specific stimuli for other insects have not been immediately successful; for example, the oligophagous tobacco hornworm does not respond behaviorally to the most likely plant-specific chemicals, the solanaceous alkaloids. The only single chemicals stimulating feeding are sucrose and inositol, which are much too ubiquitous to be the basis of host-plant specificity. The second approach to ascertaining the chemical basis of host specificity is to extract chemically and to fractionate the active principle in host-plant leaves. The chloroform fraction of the nonaqueous extract of tomato leaves has been shown to be highly stimulating for the tobacco hornworm, but further fractionation and identification of the active components are needed.

To complement this knowledge of the caterpillar's external environment, studies are in progress on the physiological basis of feeding behavior. Electrophysiological recordings of the sensory cells of caterpillars should provide a window into the nervous system to determine the basis for feeding decisions. Recordings show responses to various chemicals and plants, but the message is complex. Techniques are being developed to analyze the data by computer with the expectation of eventually understanding the sensory coding.

## Acknowledgments

I wish to thank G. deBoer, D. Sorrels, D. Klein, C. Cearley, S. Kogge, L. Schoonhoven, and K. Saxena for access to unpublished work, and G. deBoer for criticizing the manuscript.

## REFERENCES

Ahmad, S. (1982). Host location by the Japanese beetle: Evidence for a key role for olfaction in a highly polyphagous insect. *J. Exp. Zool.* **220,** 117–120.

Ave, D. A. (1981). "Induction of Changes in the Gustatory Response by Individual Secondary Plant Compounds in Larvae of *Heliothis zea* (Boddie) (Lepidoptera, Noctuidae)." Ph.D. Dissertation, Mississippi State Univ., Mississippi State.

Calvert, W. H., and Hanson, F. E. The role of sensory structures and pre-oviposition behavior in oviposition by the patch butterfly, *Chlosyne lacinia* (Geyer) (Lepidoptera: Nymphalidae). *Entomol. Exp. Appl.,* in press.

Chew, F. S. (1980). Foodplant preference of *Pieris* caterpillars (Lepidoptera). *Oecologia* **46,** 347–353.

David, W. A. L., and Gardiner, B. O. (1966). Mustard oil glucosides as feeding stimulants for *Pieris brassicae* larvae in a semi-synthetic diet. *Entomol. Exp. Appl.* **9,** 247–255.

deBoer, G., and Hanson, F. E. Chemical isolation of feeding stimulants and deterrents from tomato for the tobacco hornworm. *Proc.* Fifth Int. Symp. Insect-Plant Relationship, pp. 371–372. Pudoc, Wageningen, the Netherlands.

deBoer, G., Dethier, V. G., and Schoonhoven, L. M. (1977). Chemoreceptors in the preoral cavity of the tobacco hornworm, *Manduca sexta,* and their possible function in feeding behavior. *Entomol. Exp. Appl.* **21,** 287–298.

Dethier, V. G. (1953). Host plant reception in phytophagous insects. *Proc. Int. Congr. Entomol., 9th* **2,** 81–88.

Dethier, V. G. (1973). Electrophysiological studies of gustation in lepidopterous larvae. II. Taste spectra in relation to food–plant discrimination. *J. Comp. Physiol* **82,** 103–134.

Dethier, V. G., and Crnjar, R. M. (1982). Candidate codes in the gustatory system of caterpillars. *J. Gen. Physiol.* **79,** 549–569.

Dethier, V. G., and Kuch, J. H. (1971). Electrophysiological studies of gustation in lepidopterous larvae. *Z. Vgl. Physiol.* **72,** 343–363.

Fraenkel, G. (1959). The *raison d'etre* of secondary plant substances. *Science (Washington, D.C.)* **129,** 1466–1470.

Fraenkel, G. (1969). Evaluation of our thoughts on secondary plant substances. *Entomol. Exp. Appl.* **12,** 473–486.

Gerstein, G. L., and Clark, W. A. (1964). Simultaneous studies of firing patterns in several neurons. *Science (Washington, D.C.)* **143,** 1325–1327.

Greenblatt, J. A., Calvert, W. H., and Barbosa, P. (1978). Larval feeding preferences and inducibility in the fall webworm, *Hyphantria cunea. Ann. Entomol. Soc. Am.* **71,** 605–606.

Hamilton, S., and Hamilton, R. J. (1972). Plant waxes. *In* "Topics in Lipid Chemistry" (F. D. Gunstone, ed.), pp. 199–266. Wiley, New York.

Hanson, F. E. (1970). Sensory responses of phytophagous lepidoptera to chemical and tactile stimuli. *In* "Control of Insect Behavior by Natural Products" (D. L. Wood, R. M. Silverstein, and M. Nakajima, eds.), pp. 81–91. Academic Press, New York.

Hanson, F. E. (1976). Comparative studies on induction of food choice preferences in lepidopterous larvae. *Symp. Biol. Hung.* **16,** 71–77.

Hanson, F. E., and Dethier, V. G. (1973). Role of gustation and olfaction in food plant discrimination in the tobacco hornworm, *Manduca sexta. J. Insect Physiol.* **19,** 1019–1034.

Hanson, F., Cearley, C., and Kogge, S. Analysis of gustatory activity using computer techniques. *Proc.* Fifth Int. Symp. Insect-Plant Relationships, pp. 383–384. Pudoc, Wageningen, the Netherlands.

Howard, G. R. (1977). "Isolation of Fractions from Jimsonweed (*Datura stramonium*) and Horsenettle (*Solanum carolinense*) which Contain Hostplant Specificity Factors for the Tobacco Hornworm (*Manduca sexta*)." Ph.D. Dissertation, North Carolina State Univ., Raleigh.

Jermy, T. (1961). On the nature of the oligophagy in *Leptinotarsa decemlineata* Say (Coleoptera: Chrysomelidae). *Acta Zool. Acad. Sci. Hung.* **7,** 119–132.

Jermy, T., Hanson, F. E., and Dethier, V. G. (1968). Induction of specific food preference in lepidopterous larvae. *Entomol. Exp. Appl.* **11,** 211–230.

Johansson, A. S. (1951). The food plant preference of the larvae of *Pieris brassicae* L. *Nor. Entomol. Tidsskr.* **B8,** 187–195.

Ma, W. C. (1972). Dynamics of feeding responses in *Pieris brassicae* (Linn.) as a function of chemosensory input: A behavioral, ultrastructural and electrophysiological study. *Meded. Landbouwhogesch. Wageningen* **72-11,** 1–162.

Mordue-Luntz, A. J. (1979). The role of the maxillary and labial palps in the feeding behavior of *Schistocerca gregaria. Entomol. Exp. Appl.* **25,** 279–288.

Reese, C. J. C. (1969). Chemoreceptor specificity associated with choice of feeding site by the beetle *Chrysolina brunsvicensis* on its foodplant, *Hypericum hirsutum. Entomol. Exp. Appl.* **12,** 565–583.

Saxena, K. N., and Schoonhoven, L. M. (1978). Induction of orientational and feeding preferences in *Manduca sexta* larvae for an artificial diet containing citral. *Entomol. Exp. Appl.* **23,** 72–78.

Saxena, K. N., and Schoonhoven, L. M. (1982). Induction of orientational and feeding preferences in *Manduca sexta* larvae for different food sources. *Entomol. Exp. Appl.* **32,** 173–180.

Saxena, K. N., Khattar, P., and Goyal, S. (1976). Measurement of orientation responses of caterpillars indoors and outdoors on a grid. *Experienta* **33,** 1312–1313.

Schoonhoven, L. M. (1969). Sensitivity changes in some insect chemoreceptors and their effect on food selection behavior. *Proc. K. Ned. Akad. Wet. Ser. C* **72,** 491–498.

Schoonhoven, L. M. (1973). Plant recognition by lepidopterous larvae. *Symp. R. Entomol. Soc. London* **6,** 87–89.

Stadler, E., and Hanson, F. E. (1975). Olfactory capabilities of the "gustatory" chemoreceptors of the tobacco hornworm larvae. *J. Comp. Physiol.* **104,** 97–102.

Stadler, E., and Hanson, F. E. (1976). Influence of induction of host preference on chemoreception of *Manduca sexta:* behavioral and electrophysiological studies. *Symp. Biol. Hung.* **16,** 267–273.

Stadler, E., and Hanson, F. E. (1978). Food discrimination and induction of preference for artificial diets in the tobacco hornworm, *Manduca sexta. Physiol. Entomol.* **3,** 121–133.

Ting, A. Y. (1970). "The Induction of Feeding Preference in the Butterfly *Chlosyne lacinia.*" M. Sc. Thesis, Univ. Texas, Austin.

Waldbauer, G. P. (1962). The growth and reproduction of maxillectomized tobacco hornworms feeding on normally rejected non-solanaceous plants. *Entomol. Exp. Appl.* **5,** 147–158.

Waldbauer, G. P. (1964). The consumption, digestion and utilization of solanaceous and non-solanaceous plants by larvae of the tobacco hornworm. *Protoparce sexta* (Johan.) (Lepidoptera: Sphingidae). *Entomol. Exp. Appl.* **7,** 253–269.

Waldbauer, G. P., and Fraenkel, G. (1961). Feeding on normally rejected plants by maxillectomized larvae of the tobacco hornworm, *Protoparce sexta* (Lepidoptera: Sphingidae). *Ann. Entomol. Soc. Am.* **54,** 477–485.

Wasserman, S. S. Gypsy moth: Induced feeding preferences as a bioassay for phenetic similarity among host plants. *Proc.* Fifth Int. Symp. Insect-Plant Relationships, pp. 261–267. Pudoc, Wageningen, the Netherlands.

Wiseman, B. R., and McMillan, W. W. (1980). Feeding preferences of *Heleothis zea* larvae preconditioned to several host crops. *J. Georgia Entomol. Soc.* **15,** 449–453.

Yamamoto, R. T. (1957). "The Specificity of the Tobacco Hornworm, *Protoparce sexta* (Johan.) to Solanaceous Plants." Ph.D. Dissertation, Univ. of Illinois, Urbana.

Yamamoto, R. T. (1974). Induction of hostplant specificity in the tobacco hornworm, *Manduca sexta. J. Insect Physiol.* **20,** 641–650.

Yamamoto, R. T., and Fraenkel, G. (1960a). Assay of the principle gustatory stimulant for the tobacco hornworm, *Protoparce sexta,* from Solanaceous plants. *Ann. Entomol. Soc. Am.* **53,** 499–503.

Yamamoto, R. T., and Fraenkel, G. (1960b). The physiological basis for the selection of plants for egg-laying in the tobacco hornworm, *Protoparce sexta* (Johan.) *Proc. Int. Cong. Entomol., 11th* **3,** 127–133.

Yamamoto, R. T., and Fraenkel, G. (1960c). The specificity of the tobacco hornworm, *Protoparce sexta* Johan.), to solanaceous plants. *Ann. Entomol. Soc. Am.* **53,** 503–507.

# PART II

# The Diversity of Behavioral Cues

# 2

# Chemical Aspects of Oviposition Behavior in Butterflies*

PAUL FEENY, LORRAINE ROSENBERRY, AND MAUREEN CARTER



## I. INTRODUCTION

Because of the vagaries of their evolutionary history, the larvae of most butterfly species grow and survive best on only a limited number of host-plant species often belonging to a single family. Because the tiny hatchling larvae are usually incapable of moving great distances in search of food, they are dependent to a considerable extent on the host-finding skills of their mothers. Even when provided with a suitable start in life, few offspring survive to become adults.

*This work has been supported by research grants DEB 76-20114 and DEB 79-22130 from the National Science Foundation.



HERBIVOROUS INSECTS

ISBN 0-12-045580-3

Most fall victim to predators, parasitoids, or pathogens, to physical catastrophes such as drowning and desiccation, or to starvation when their local food supply is exhausted (Gilbert and Singer, 1975). Adult females of most butterfly species seldom live longer than a few weeks, and often for only a few days. To have any chance of endowing the next generation with some of her own progeny, a female must lay her eggs rapidly as well as accurately. It comes as no surprise to find that butterflies possess some elaborate behavioral and physiological mechanisms to help them in their quest for suitable oviposition sites.

The earliest published suggestion that chemistry plays a role in oviposition behavior by butterflies seems to have been that of Verschaffelt (1911). He noted that larval food plants of the cabbage white butterflies *Pieris brassicae* and *P. rapae,* although belonging to different plant families, share the presence of glucosinolates in their tissues. Plants lacking these compounds were at best nibbled on by *Pieris* larvae, though some became acceptable when smeared with leaf juices from a cruciferous plant or with solutions of sinigrin (allylglucosinolate). Because the nonvolatile glucosinolates give rise to volatile fission products (primarily isothiocyanates, the "mustard oils"), Verschaffelt could not be sure of the extent to which the larvae were responding to odor as opposed to taste, though "since caterpillars do not as a rule taste other plants, it is fairly clear that they must be informed as to the nature of the food offered by its odour." As to the egg-laying adults, he was less doubtful: "This is to a still greater extent necessary for the butterflies, which only lay their eggs on Crucifers or on plants chemically related to them, and will doubtless recognise these by the odour [Verschaffelt, 1911, p. 539]."

Since Verschaffelt's original publication, the role of chemistry in the oviposition behavior of butterflies has become better defined as a result of extensive phytochemical comparisons among larval host plants and of intensive studies of sensory physiology and oviposition behavior. This chapter reviews some general contributions from these complementary lines of study and considers in more detail some applications of both approaches to the study of oviposition behavior in butterflies of a single family, the Papilionidae.

## II. HOST RELATIONSHIPS AMONG BUTTERFLIES

Brues (1920) had already realized that the oviposition behavior of female butterflies and moths provided a potential key to understanding the evolution of host-plant patterns among the Lepidoptera. Like Verschaffelt (1911), he considered that food plants are selected on the basis of odor by the parent female and also selected on the same basis by the larvae. Taste, "which is no doubt closely connected with odor [p. 332]," was also believed to play a role, at least for the larvae. After noting that the larval feeding habits of several groups of butterflies have remained conservative over enormous periods of time, Brues (1920) dismissed the idea that changes in host-use patterns result merely from the influence on adults of larval experience on novel food plants. He went on to say: "We can more easily believe that they may have arisen through mutations in maternal instinct not incompatible with larval tastes and then only in extremely rare cases [p. 329]."

Although well aware of the chemical similarities noted by Verschaffelt among even unrelated food plants of *Pieris* butterflies, Brues (1920) found it hard to believe more generally that "oligophagous or polyphagous species have become accustomed to a variety of plants due to a confusion of similar odors [p. 331]." It fell to Dethier (1941) to point out that such chemical facilitation of changes in feeding habits has probably been common and that the evolution of host relationships among the butterflies, at least, has not depended solely on the kind of sudden major mutations implied by Brues. Dethier showed experimentally that larvae of the black swallowtail butterfly, *Papilio polyxenes* (then called *P. ajax*), are attracted by the odors of several of the essential oils that are found in the umbelliferous food plants of this species. He noted that *P. polyxenes* occasionally feeds on certain plants in the Rutaceae, a family fed on extensively by many other species of *Papilio*. Because several of the attractive essential oils occur variously in the Rutaceae, Dethier suggested an evolutionary sequence from Rutaceae-feeding to Umbelliferae-feeding swallowtails on the basis of the role of particular chemical compounds as catalysts of evolutionary change: "The transition from one plant family to the other took place because of the presence of identical attractant chemicals in both families [Dethier, 1941, p. 72]."

Although Dethier's experiments and discussion were restricted to larvae, he says: "There is no escaping the fact that the choice of food is largely predetermined by the gravid female at the time she selects the place to lay her eggs [Dethier, 1941, p. 61]." One is left with the implication that the proposed evolutionary scheme may have resulted as much from the attractiveness of certain compounds to the ovipositing females as from their attractiveness to the larvae. The assumption of Verschaffelt (1911) and Brues (1920) that similar compounds are used as host-finding cues by females and larvae of a species eventually received experimental support when David and Gardiner (1962) showed that females of *Pieris brassicae* would lay eggs readily on pieces of green paper that had been treated with aqueous solutions of allylglucosinolate.

Ehrlich and Raven (1964) greatly extended the observations of Verschaffelt (1911) and Dethier (1941) by showing that chemical similarity among the larval food plants of related butterflies is a widespread phenomenon and frequently cuts across botanical affinities. Depending on the butterfly group in question, correlations of feeding habits were found with a variety of different natural products, including alkaloids, coumarins, essential oils, glucosinolates, acetylenic fatty acids, and bitter-tasting phenolic glycosides. Ehrlich and Raven (1964), following Fraenkel (1959), emphasized the role of these compounds as physiological barriers, variously overcome by butterfly larvae in the course of their evolution. In conjunction with the findings of David and Gardiner (1962), however, their documentation of widespread correlations between chemical patterns and the larval feeding habits of butterflies provided circumstantial support for the hypothesis that unique classes of secondary compounds are more generally involved as "sign stimuli" in the oviposition behavior of butterflies.

## III. CHEMOSENSORY SYSTEMS IN BUTTERFLIES

When searching for plants on which to lay their eggs, female butterflies of many species have been observed to move their forelegs vigorously back and forth in rapid alternation on the surface (usually a leaf) of almost every plant on which they alight (Ilse, 1937a). During this "drumming reaction," as it was called by Ilse (1937b), the terminal tarsal segments are brought sharply, and sometimes audibly, into contact with the leaf and scraped along the surface at the end of each stroke (Fox, 1966). Drumming is generally followed by takeoff and resumption of searching flight unless the plant being drummed upon belongs to one of the larval host species. In this case, drumming may be followed by oviposition, either on the same leaf or nearby. After studying drumming behavior in a variety of butterfly species, Ilse (1956) concluded that it must provide information as to the physical and chemical nature of leaves, thus permitting females to distinguish among plant species.

A corollary of Ilse's hypothesis (1956) is that female butterflies must possess sensory receptors on their forelegs. Convincing evidence for the presence of chemoreceptors on the legs of butterflies had in fact been obtained much earlier by Minnich (1921). When he brought the tarsi of any of the four walking legs of male or female red admiral butterflies (*Vanessa atalanta;* Nymphalidae) into contact with apple juice or sucrose solutions, the butterflies instantly uncoiled their proboscises in an apparent feeding response. Among butterflies deprived of water for several days, a similar response could be elicited by distilled water. Removal of antennae, labial palps, and rudimentary forelegs scarcely affected the frequency of proboscis extension as long as the legs were brought into contact with the test solution. Responses to apple juice or sucrose could not be obtained, however, when the ventral surfaces of the tarsi were held as close as 1 mm from solutions known to be stimulatory on contact. Minnich (1921) concluded that the butterflies must possess contact chemoreceptors on the tarsi of all four walking legs and thus that butterflies can "taste with their feet." Further experiments by Minnich (1921, 1922a,b), Weis (1930), Anderson (1932), and Frings and Frings (1949, 1956) showed that tarsal contact receptors sensitive to sucrose are of widespread and perhaps general occurrence among butterflies.

Eltringham (1933) examined the tarsi of male red admiral butterflies and found thin-walled trichoid sensilla, set in sockets and spaced along the ventral surface. Grabowski and Dethier (1954) found similar hairs, thin-walled, blunt-tipped, and usually curved, on the mid- and hindlegs of butterflies from three families and of one skipper species. As was found by Eltringham (1933), the sensilla were scattered singly and did not seem to differ appreciably between the sexes. The supposition that these sensilla include the contact receptors responsible for the detection of sugars and also of salt solutions has since been amply confirmed by electrophysiological techniques (Morita and Takeda, 1957, 1959; Morita *et al.*, 1957; Takeda, 1961).

The individual contact receptors responding to water, sucrose, and salt solutions could play a supplementary role in oviposition behavior, but their presence on all walking legs of both male and female butterflies, together with their association with proboscis extension, have long been taken as evidence that their primary function is related to feeding. Fox (1966) was evidently the first to concentrate attention on the sensory structures of the forelegs of females. He examined several hundred butterflies, belonging to all families, and found clusters of trichoid sensilla on the ventral surfaces of the tarsal segments (tarsomeres) of the forelegs of all females. These clusters were absent from the forelegs of males and from the mid- and hindlegs of both sexes.

Because sucrose sensitivity is absent from the foretarsi of nymphalids, Fox (1966) suspected that these clustered sensilla must serve some additional function. Further evidence for this came from an examination of the tarsi of males and females of *Pieris rapae,* in which sucrose sensitivity is known to be present in the fore- and midlegs of both sexes but absent from the hindlegs (Frings and Frings, 1949). Fox (1966) found that this correlated well with the distribution of scattered sensilla but left unexplained the function of the clustered sensilla that he found only on the foretarsi of females. Aware of Ilse's description (1937b) of drumming behavior but apparently unaware of her later hypothesis as to its function (Ilse, 1956), Fox (1966) suggested independently that the function of drumming is to provide chemoreceptive information as to food-plant identity, the chemoreceptors involved being represented by the clusters of sensilla unique to the forelegs of females.

Fox (1966) noted that the clustered sensilla on the foretarsi of females were generally associated with spines. On the foretarsi of females of *Speyeria cybele,* which he examined in particular detail, tight clusters of 4–12 sensilla occurred in pairs on each of the four distal tarsal segments. Each pair of clusters, located at the proximal end of its tarsomere, was intimately associated with a pair of distal-pointing spines located at the distal end of the preceding (more proximal) tarsomere. The close association of clumped sensilla with spines (see Fig. 1) suggested to Fox (1966) that the spines serve to abrade leaf surfaces during drumming, releasing essential oils that are then detected by the clumped sensilla.

**Fig. 1.** (a) Part of the ventral surface of the fifth tarsomere of a foreleg of a female black swallowtail butterfly, *Papilio polyxenes*. Clumped sensilla, associated with the central grooved spines, are likely sites for contact chemoreception during oviposition behavior. Smaller isolated sensilla can also be seen among the field of microtrichia. (b) Enlargement of one of the central spines, indicating close association with tips of clumped sensilla. (c) Female of *P. polyxenes* showing abdomen-curling response after alighting on a sponge leaf during a free-flight bioassay experiment. (d) Female of *P. polyxenes* curling her abdomen during a filter-paper bioassay. (Photographs a and b by E. R. Hoebeke.)

Pairs of sharply pointed spines, associated with clusters of sensilla, were also found by Calvert (1974) on the ventral surfaces of the first four tarsal segments of the (reduced) forelegs of females of *Chlosyne lacinia* (Nymphalidae). Calvert (1974) noticed that the spines in *C. lacinia* were markedly grooved and that the trichoid sensilla twisted around each other and were aligned along or lay within the lateral groove of the corresponding spine (see Fig. 1). He suggested that if the spines serve to puncture the surfaces of leaves, as suggested by Fox (1966), the grooves in the spines might serve to collect leaf juices, thus permitting them to be more readily sampled by the trichoid sensilla.

Myers (1969) tested the role of foretarsi in oviposition behavior by means of ablation experiments with Florida queen butterflies, *Danaus gilippus berenice*. She inactivated any antennal chemoreceptors by coating both antennae with paint and then removed various combinations of tarsi from the legs. When tarsi were removed from all six legs, the females laid no eggs on *Asclepias* plants. When the foretarsi were left intact, however, the females laid eggs even if the tarsi had been removed from all four walking legs. The foretarsi are thus sufficient for oviposition. When just the foretarsi were removed, however, the females would still lay eggs, suggesting that appropriate receptors are not restricted to the foretarsi and that receptors on the tarsi of other legs may serve as a "backup system" (Myers, 1969).

Ma and Schoonhoven (1973) combined behavioral and electrophysiological experiments in a thorough investigation of the role of tarsal chemoreceptors in the oviposition behavior of *Pieris brassicae*. They caged mated *P. brassicae* females with freshly cut broad bean plants, *Vicia faba*, that had been cultured in aqueous solutions of glucosinolates. The females laid eggs freely on these plants, which are not normally used as oviposition or larval food plants, and they laid no eggs on adjacent control plants that had been cultured in water. Subsequent ablation experiments by Ma and Schoonhoven (1973) yielded results comparable to those obtained by Myers with queen butterflies. The oviposition response of *P. brassicae* females to the experimental plants was not inhibited by removing the antennae, nor was it diminished by removal of the forelegs or inactivation of the mid- and hindlegs. In fact, females lacking their foretarsi laid eggs at the same rate as did the control females. Only when the tarsi of all six legs were inactivated was oviposition suppressed completely. The glucosinolate receptors must therefore be located on the tarsi of the legs, but not merely on those of the forelegs.

Ma and Schoonhoven (1973) next examined the tarsi of both sexes of *Pieris brassicae* and found two categories of cuticular projections in addition to scales and microtrichia. The first category, Type A, included straight, stout, tapering setae, and also spines and bristles of various lengths. Some of the latter were innervated by a bipolar sense cell and were considered to be tactile receptors.

Smaller and more delicate were the Type B hairs, which were slightly curved, blunt-tipped, and based in membranous sockets. Type A and Type B structures were present on the tarsi of both males and females, B hairs being more numerous in females. On the fifth (distal) tarsomere of the foretarsus, where sexual differences were most marked, the B hairs were grouped into a pair of clusters, each containing 13–15 hairs in females and 6–7 hairs in males.

Examination of B hairs on the fifth tarsomeres of females revealed that each hair was associated with five bipolar sense cells. Distal processes from four of the cells extended into the hair itself, where they were surrounded by a fluid-filled space (the trichogen cell) that communicated with the external environment by means of a small pore at the hair tip. Stimulation of a hair with NaCl solutions produced two types of electrophysiological responses, corresponding to "S1" and "S2" cells. No additional response was obtained after stimulation with sucrose solutions, but stimulation with glucosinolate solutions produced a new kind of spike (corresponding to the "G" cell), whose discharge frequency (unlike those of the S1 and S2 cells) increased with the concentration of the stimulating solution. Even slight deflections of the hairs were found to produce discharges from a very sensitive mechanoreceptor cell, probably the cell whose dendrite does not extend far into the hair lumen.

Because the small number of receptor cells associated with the B hairs corresponds with the typical number for other taste sensilla in insects—whereas olfactory sensilla are generally associated with many sensory cells—and because the electrophysiological experiments showed that four of the cells respond to stimulation by solutions, Ma and Schoonhoven (1973) concluded that the B hairs contain, in addition to the mechanoreceptor cell, four cells that serve as contact chemoreceptors. One probably responds to water, another to salt, and at least one other to glucosinolate solutions, which are detectable down to molar concentrations of $10^{-5}$–$10^{-4}$. They suggest two functions for the B hairs, namely, the discrimination of salt solutions from water and the detection of "the chemical principle in the complex of stimuli inducing oviposition [p. 354]."

Minnich (1921) showed long ago that red admiral butterflies, *Vanessa atalanta,* can detect the presence of apple juice by olfaction as well as by taste. In contrast to the many subsequent studies of contact chemoreception, however, olfactory sense organs in butterflies seem to have been virtually ignored. In the only detailed study of which we are aware, Behan and Schoonhoven (1978) identified four types of sensilla on the antennae of male and female *Pieris brassicae* butterflies. The two most abundant types were considered to have an olfactory function. Electroantennograms and recordings from single sensilla showed that the antennal receptors can distinguish among a wide range of volatile plant compounds and can detect an aversion pheromone associated with conspecific eggs (see Rothschild and Schoonhoven, 1977).

# IV. STUDIES OF SWALLOWTAIL BUTTERFLIES

## A. Introduction

Although the role of chemistry in the oviposition behavior of butterflies is probably best understood among species of the family Pieridae, considerable attention has also been devoted to several species within the related family Papilionidae. These studies have been undertaken partly for economic reasons, because several swallowtail species are pests of citrus and umbellifer crops. In addition, however, the family Papilionidae has several characteristics that render it an ideal group in which to explore more generally the processes that have shaped the evolution and adaptive radiation of a group of herbivorous insects.

The swallowtail family comprises about 530 species, currently arranged within three subfamilies and six tribes (Fig. 2) (Munroe, 1960; Munroe and Ehrlich, 1960). Although relatively small, the family includes species that inhabit a wide range of environments, including temperate deciduous forests, boreal forests, deserts, savannas, arctic and alpine tundra, prairies, freshwater marshes, and tropical rain forests. The larvae of some species feed on herbaceous plants, those of others on shrubs or trees; most are oligophagous, but a few are widely polyphagous. Despite the ecological diversity of the group, striking patterns of chemical similarity among its larval food plants suggest that behavioral and/or toxicological adaptation to various classes of compounds has restricted some directions of evolutionary change while facilitating others (Dethier, 1941, 1954; Ehrlich and Raven, 1964).

Of the three tribes in the largest subfamily, Papilioninae, the tribe Troidini consists of species that feed, as larvae, primarily on plants of the family Aristolochiaceae (Table I). The species of the tribe Graphiini feed chiefly on families in the order Magnoliales (subclass Magnoliidae) (Cronquist, 1968), including the Annonaceae, Lauraceae, and Magnoliaceae. The third tribe, Papilionini, consists only of the large genus *Papilio,* most species of which are restricted to the Rutaceae, Lauraceae, or related families. Members of the holarctic *Papilio machaon* group, however, feed mostly on the Umbelliferae, but a few species feed on the Asteraceae (genus *Artemisia*) and Rutaceae (Ehrlich and Raven, 1964; Scriber, 1973).

Munroe and Ehrlich (1960) and Ehrlich and Raven (1964) have argued reasonably, on morphological as well as on chemical grounds, that the Aristolochiaceae feeders represent remnants of the stock from which the rest of the Papilionidae evolved (with the probable exception of *Baronia brevicornis,* the only species in the most primitive subfamily Baroniinae). The families of the order Magnoliales (subclass Magnoliidae) are linked by their content of benzylisoquinoline alkaloids both with the Aristolochiaceae and with the Rutaceae (Table II; Fig. 3).

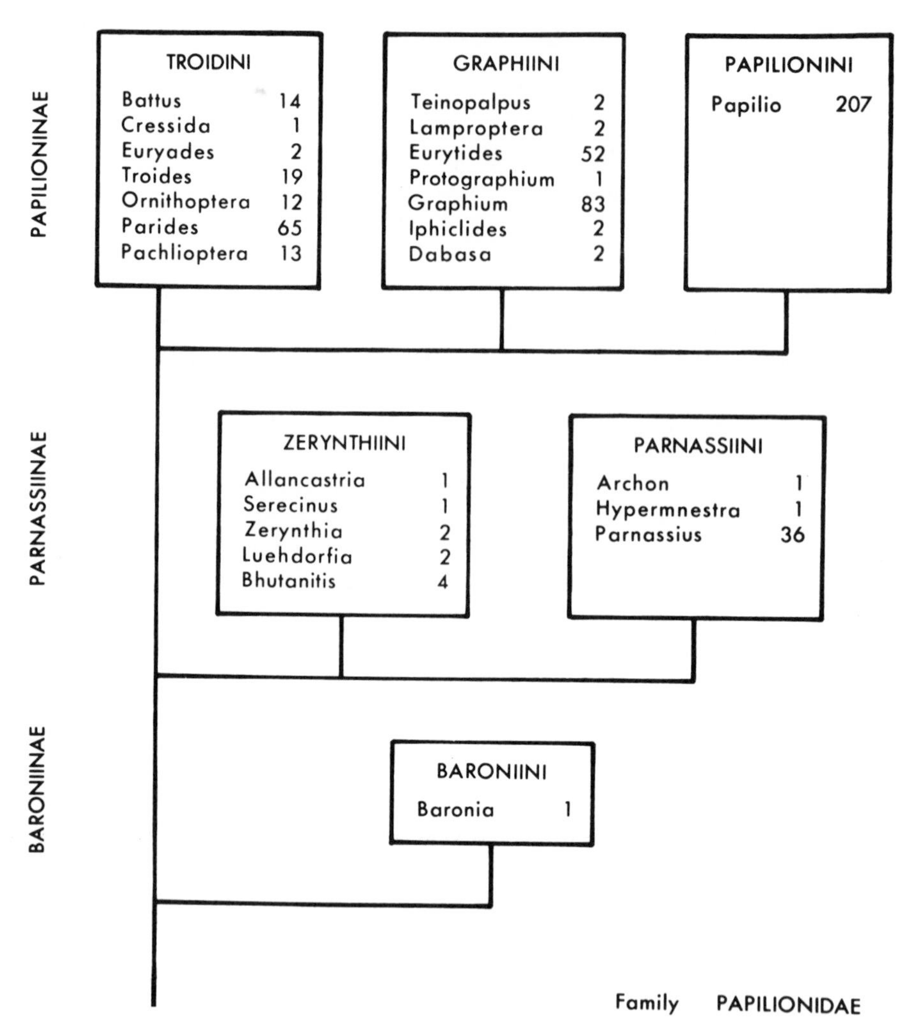


**Fig. 2.** The genera and numbers of species of swallowtail butterflies, arranged vertically by subfamilies and horizontally by tribes. (Data from Munroe, 1960 and Munroe and Ehrlich, 1960.)

Hydroxycoumarins and furanocoumarins are likewise associated with larval food-plant families, especially of the supposedly more recent genus *Papilio*. The most consistent chemical link of all is provided by the essential oils, which are reported from virtually all families of swallowtail food plants (Table II; Fig. 3). The ethereal oil cells in the Rutaceae are morphologically similar to those in the Aristolochiaceae, Piperaceae, and families of the order Magnoliales (subclass Magnoliidae) (Cronquist, 1968). Essential oils found in the Rutaceae and Um-

**TABLE I.** Numbers of Swallowtail Species Associated with the Major Families of Larval Food Plants[a]

| Major Food-Plant Families | Tribes of the Family Papilionidae | | | | | |
|---|---|---|---|---|---|---|
| | Baroniini | Zerynthiini | Parnassiini | Troidini | Graphiini | Papilionini |
| Subclass Magnoliidae | | | | | | |
| Aristolochiaceae | — | 10 | 1 | 70 | — | — |
| Fumariaceae | — | — | 3 | — | — | — |
| Piperaceae | — | — | — | 1 | — | 4 |
| Annonaceae | — | — | — | — | 32 | — |
| Lauraceae | — | — | — | — | 8 | 18 |
| Magnoliaceae | — | — | — | — | 4 | 5 |
| Subclass Rosidae | | | | | | |
| Leguminosae | 1 | — | — | — | — | 1 |
| Crassulaceae | — | — | 4 | — | — | — |
| Saxifragaceae | — | — | 3 | — | — | — |
| Rosaceae | — | — | — | 1 | 2 | 4 |
| Rutaceae | — | — | — | 8 | 3 | 74 |
| Rhamnaceae | — | — | — | — | — | 3 |
| Umbelliferae | — | — | — | — | — | 16 |
| Subclass Asteridae | | | | | | |
| Verbenaceae | — | — | — | — | 3 | 1 |
| Rubiaceae | — | — | — | — | — | 4 |
| Asteraceae | — | — | — | — | — | 4 |
| Subclass Hamamelidae | | | | | | |
| Betulaceae | — | — | — | — | — | 3 |

[a]Data from Scriber (1973) and Tyler (1975). Except for Leguminosae, only those families are listed that contain species used as food plants by at least three swallowtail species.

belliferae are known to act as larval attractants both in *Papilio polyxenes* (Dethier, 1941) and in the lemon butterfly, *Papilio demoleus* (Saxena and Prabha, 1975).

Within the genus *Papilio,* especially among species associated with the Rutaceae and Umbelliferae, there have been numerous observations, both in the field and in the laboratory, of females of one species laying eggs on host plants used more typically by the larvae of other species (e.g., Dethier, 1941; Stride and Straatman, 1962; Abe *et al.*, 1981; Berenbaum, 1981a). Additionally, Stride and Straatman (1962) found that females of *P. aegeus,* a Rutaceae feeder, would lay eggs on *Cinnamomum camphora* (Lauraceae), a larval food plant of *Graphium* species, and D. A. West (personal communication) found an egg of the spicebush swallowtail, *P. troilus,* on a leaf of *Aristolochia macrophylla.* Such observations suggest an underlying conservatism in chemical adaptation among swallowtails as a whole. It seems necessary only to identify the key class of

**TABLE II.** Distribution Pattern of Typical Classes of Secondary Compounds Found in Families of Plants Most Used as Larval Food by Swallowtail Species[a]

| Major Food-Plant Families | Essential Oils | | Benzylisoquinoline Alkaloids | Hydroxycoumarins | Furanocoumarins |
|---|---|---|---|---|---|
| | Terpenoid | Phenylpropanoid | | | |
| Subclass Magnoliidae | | | | | |
| Aristolochiaceae | X | X | X | — | — |
| Fumariaceae | — | — | X | — | — |
| Piperaceae | X | X | — | — | — |
| Annonaceae | X | — | X | — | — |
| Lauraceae | X | X | X | — | — |
| Magnoliaceae | X | X | X | X | — |
| Subclass Rosidae | | | | | |
| Leguminosae | X | X | ? | X | X |
| Crassulaceae | ?[b] | — | — | — | — |
| Saxifragaceae | X | — | — | X | — |
| Rosaceae | — | ? | — | X | — |
| Rutaceae | X | X | X | X | X |
| Rhamnaceae | ? | ? | X | — | — |
| Umbelliferae | X | X | — | X | X |
| Subclass Asteridae | | | | | |
| Verbenaceae | X | — | — | — | — |
| Rubiaceae | X | X | — | X | — |
| Asteraceae | X | X | — | X | — |
| Subclass Hamamelidae | | | | | |
| Betulaceae | X | — | — | — | — |

[a]Data primarily from Guenther (1948–1952), Hegnauer (1964–1973), Heywood (1971), Robinson (1975), Seigler (1977), and Berenbaum (1980).
[b]Question mark indicates unconfirmed reports of occurrence.

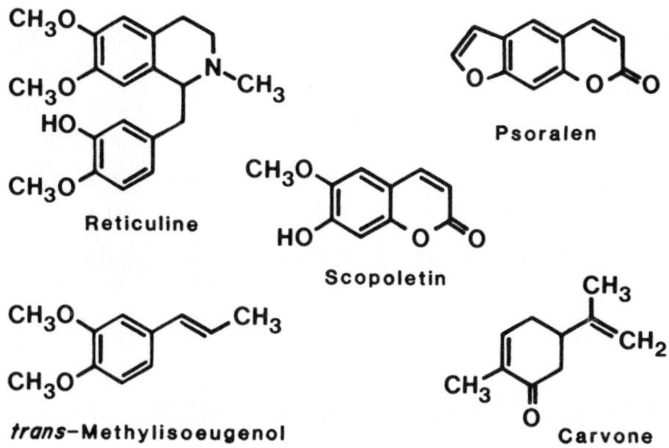

**Fig. 3.** Structural formulas for examples of the classes of secondary compounds most characteristic of swallowtail food-plant families: reticuline (a benzylisoquinoline alkaloid), scopoletin (a hydroxycoumarin); psoralen (a furanocoumarin), *trans*-methylisoeugenol (a phenylpropanoid essential oil), and carvone (a terpenoid essential oil).

compounds used as sign stimuli for oviposition, analogous to the glucosinolates in food plants of the pierids, and the door to understanding the evolution of the Papilionidae will be unlocked. This endeavor has not proved to be simple, however, and its goal remains elusive.

## B. The Tribe Troidini

Vision plays the predominant role in the plant-search phase of oviposition behavior by females of the pipevine swallowtail, *Battus philenor* (Rausher, 1978). In the longleaf pine savannas of southeastern Texas, females search the open understory for the larval food plants, *Aristolochia reticulata* and *A. serpentaria,* by forming search images for the shapes of the leaves. Because a particular leaf shape is shared by many other plant species, the females typically land on many nonhost plants before they finally alight on the sought-for *Aristolochia,* which is presumably then recognized by contact chemoreception.

Nishida (1977) investigated the role of contact oviposition stimuli in oviposition behavior by *Byasa (= Parides) alcinous,* the larvae of which feed on *Aristolochia debilis*. He found that when females of this species were placed on filter paper treated with methanolic extracts of *A. debilis* roots they would drum with their forelegs, curl their abdomens, and lay one or more eggs on the paper. Removal of both antennae did not reduce this behavior, but coating the fourth and fifth tarsomeres of the forelegs with nail polish inhibited the response completely. Nishida (1977) attributed the oviposition response to a trichoid organ visible on the fifth tarsomere of females but apparently absent in males.

In an attempt to isolate and identify the compounds responsible for activity in the methanolic extracts of *A. debilis* roots, Nishida (1977) subjected the extract to steam distillation, followed by solvent extractions of the aqueous residue with ether and ethyl acetate. The resulting water layer, now lacking the most nonpolar compounds from the original extract, was treated with ammonium reineckate to precipitate alkaloids. Bioassay of all fractions revealed no activity in the steam distillate, in either of the organic extracts, or in the precipitate. Only the water layer remained active throughout, indicating that the oviposition stimulants are polar.

### C. The Tribe Papilionini: Rutaceae-Feeding Species

In a series of pioneering studies, Vaidya (1956, 1969a,b) and Ilse and Vaidya (1956) examined the roles of odor, color, and form perception during feeding and oviposition behavior by the lemon butterfly, *Papilio demoleus*. Vaidya (1969b) exposed mated females, with no prior experience of color, to artificial paper leaves, shaped and colored to resemble *Citrus* leaves and pinned to corks in a rectangular array on a verticle board. Provided that the odor of *Citrus* was present, females would fly to the artificial leaves and drum on the surface with their forelegs. When offered a choice of differently colored "leaves," females overwhelmingly preferred those that were colored blue-green or green, ignoring the colors found to be most attractive to feeding butterflies (Ilse and Vaidya, 1956). In the absence of *Citrus* odor, or if the odor was too concentrated, the butterflies did not discriminate among colors (Vaidya, 1969b). Though the location of its source was not important, the presence of host scent was thus vital for behavioral orientation. Vaidya (1956, 1969b) concluded that the odor of the host plant is of prime importance during oviposition by *P. demoleus* whereas color plays a secondary role. Even in the presence of odor and a blue-green surface, the drumming females would seldom lay eggs, suggesting that vision and odor are only part of the story.

Saxena and Goyal (1978) conducted further laboratory studies of oviposition behavior by Indian *Papilio demoleus* butterflies. To assess the role of various possible cues used by females when approaching plants (a behavior they called "orientation"), they placed mated females, three at a time, inside a transparent plastic box (20 × 20 × 20 cm), the left and right sides of which had been replaced with translucent nylon netting. In a series of paired choice tests, the number of visits by the females to experimental substrates placed on one side of the box were compared with the number of visits to appropriate controls on the other side. Muslin-covered filter-paper disks, colored green or yellow-green and placed vertically outside the netting, elicited strong preferences relative to white control disks, whereas disks of other colors were less attractive or even repellent. In contrast to the finding of Vaidya (1969b), the presence of host-plant odor was not required for color discrimination (Saxena and Goyal, 1978; K. N. Saxena, personal communication). Saxena and Goyal (1978) also found that the degree of preference for yellow-green disks was not enhanced by moistening them with water. In none of these experiments with colored disks, whether moistened or not, was abdomen-curling or oviposition observed on the netting.

In experiments exploring the role of host odors in orientation, Saxena and Goyal (1978) found that lime leaves, *Citrus limettioides,* placed outside the netting were preferred over similar leaves that were shielded behind a glass screen. Because visual cues were presumably equalized, the authors attributed the preference to a response to host odor. Of particular interest was the observation that almost half of the visits to the side adjacent to unscreened leaves resulted in abdomen-curling (versus 20% in visits to the side adjacent to the screened leaves), and in about a third of these cases eggs were laid on the netting of the cage. The authors next found that unmoistened muslin-covered disks, treated with an ether extract of fresh lime leaves and placed vertically outside but adjacent to the cage netting, were preferred over blank control disks, though no abdomen-curling was observed. When the disks containing ether extract were moistened, however, the preference for them was increased greatly relative to the (moistened) blank control disks, and abdomen-curling was seen in 67% of the visits (versus 13% in visits to the control side of the cage). Saxena and Goyal (1978) concluded from these and related experiments that a combination of color, moisture, and host-plant odor is required for most effective orientation toward potential oviposition sites. Their results also show that this combination suffices to stimulate oviposition, albeit at low levels, in the absence of contact with a leaf surface.

To evaluate the role of contact stimuli, Saxena and Goyal (1978) next offered *Papilio demoleus* females experimental muslin-covered disks that were placed inside the bioassay cage. A moistened disk treated with ether extract of lime leaves elicited a 75.6% abdomen-curling response by females landing upon it, followed in most cases by egg laying. This confirmed the earlier finding that some combination of moisture, nonpolar volatile compounds (ether extract), and perhaps color is sufficient to stimulate oviposition. However, a moistened disk treated with an 80% ethanol extract of (ether-extracted) lime leaves elicited a higher response (80.8% abdomen-curling, 66.1% egg laying), even though earlier tests had shown that the ethanol extract (unlike the ether extract) failed to stimulate abdomen-curling or oviposition when applied to a moistened disk outside the cage. The optimum oviposition response was obtained by applying both ether and ethanol extracts to moistened yellow-green disks. Saxena and Goyal (1978) concluded that volatile stimuli, combined with moisture and visual stimuli, are sufficient to stimulate the entire sequence of oviposition behaviors but that the responses are enhanced by contact perception of nonvolatile (more polar) components present in ethanolic extracts.

In recent field experiments, K. N. Saxena (personal communication) has found that perforated yellow-green plastic buckets, containing an ether extract of lime leaves and suspended from *Citrus* trees, are highly attractive to wild *Papilio demoleus* females. When the outsides of the buckets were roughened and coated with the sticky, concentrated ethanolic extract of young *Citrus* leaves, many more eggs were laid on the buckets than on nearby young *Citrus* leaves of greater available surface area.

A second *Citrus*-feeding swallowtail to receive detailed attention has been *Papilio protenor demetrius.* This species was studied by Ichinosé and Honda (1978) in Japan, where the larvae feed on several rutaceous tree species, including *Poncirus trifoliata, Citrus junos,* and *Fagara ailanthoides.* Egg-laying females of this species do not approach suitable host plants directly but wander from plant to plant drumming the leaf surfaces of various species, including nonhost plants. Should the leaf belong to one of the species usually used as a larval food plant, drumming is generally followed by oviposition. Should the females fail to land on a suitable leaf after several drumming events in a particular locality, however, they leave and fly to another site. The entire sequence of oviposition behaviors is accompanied by a characteristic rapid fluttering of the wings (Ichinosé and Honda, 1978; cf. Vaidya, 1956).

Ichinosé and Honda (1978) set out to test the hypothesis that essential oils are involved in the oviposition behavior of *Papilio* species. They concentrated their work on the responses that follow contact with the oviposition substrate. Females of *P. protenor* were first screened for positive oviposition responses when placed on leaves of a larval host plant and for negative responses on moistened strips of filter paper. They were then tested for their responses to various plant extracts and solutions of pure compounds, applied to similar filter-paper strips. Abdomen-curling, whether or not followed by egg laying, was taken as the criterion for a positive response and the overall response for a female, assayed 10–20 times per day at 10-min intervals, was scored simply as the number of positive responses divided by the number of trials. Aqueous extracts of leaves of three larval host plants (*Poncirus trifoliata, Fagara ailanthoides,* and *Zanthoxylum planispinum*) elicited overall responses of 0.6 to 1.0, whereas extracts of two rutaceous plants not used as larval hosts (*Phellodendron amurense* and *Orixa japonica*) produced responses of only 0 to 0.15. An extract of the epicarp of lemons, *Citrus limon,* was found to be highly active (0.90). Because females would not respond to intact lemon fruits, apparently because they could not settle on the smooth surface, the authors concluded that surface texture is important in oviposition behavior. Color, on the other hand, was considered to be unimportant because the active filter papers containing expressed juice of lemon epicarp were essentially colorless. As found by Saxena and Goyal (1978) for *Papilio demoleus,* the presence of moisture was essential for a positive contact response, whereas filter paper treated with water alone was not stimulatory.

Ichinosé and Honda (1978) found that cutting off both antennae at the base did not affect the oviposition response of *Papilio protenor* females. When both foretarsi were removed or coated with nail polish, however, there was little or no response to *Poncirus* leaves. The authors concluded that the foretarsi are important and that oviposition is stimulated by certain contact chemical(s). Examination of the tarsi revealed rows of conspicuous black spines accompanied by minute hairs on the ventral surfaces of the tarsomeres. The tarsi of the mid- and hindlegs did not differ obviously between the sexes. The fifth tarsomere of the foretarsus of females, however, was about half the length of that of the males, but much thicker. The minute hairs were more abundant than in the male, giving a dense brushlike appearance. Ichinosé and Honda (1978) concluded that these brushlike hairs in the females are trichoid sensilla, including receptors sensitive to the oviposition factor(s).

Using the bioassay technique described, Ichinosé and Honda (1978) found that *Papilio protenor* females were strongly stimulated by aqueous extracts of lemon epicarp, even after prior extraction with organic solvents, but that the organic extracts themselves (except for that in acetone) were inactive. A further indication that essential oils are probably not involved in contact stimulation of oviposition behavior was that steam distillates of host-plant leaves were also inactive. The authors finally assayed acetone solutions, at five different concentrations, of various *Citrus* oils, of (−)-carvone and (+)-carvone, and of 13 pure compounds known to occur in *Poncirus trifoliata* (citral, citronellal, linalool, nerol, myrcene, (+)-limonene, α-terpineol, α-pinene, β-pinene, *n*-decylaldehyde, *n*-nonylaldehyde, geranyl acetate, citronellyl acetate). Of all the solutions, only nerol (10%) showed any activity whatsoever; because *Citrus* oils containing nerol were inactive, this response was considered to be due to a certain abnormal state of physiology. The authors conclude that odor, like color, is unimportant in the contact phase of oviposition by *P. protenor*, though it could play a role in the earlier plant-searching behavior. They conclude that the chemical components directly responsible for contact stimulation are water-soluble and thermostable, though probably of relatively low molecular weight.

Similar conclusions were reached by Nishida (1977) after investigating the contact phase of oviposition behavior in *Papilio xuthus*, another Rutaceae-feeding swallowtail in Japan. Using a filter-paper bioassay, similar to that of Ichinosé and Honda (1978), Nishida (1977) found that females of this species responded strongly to methanolic extracts of *Citrus natsudaidai*, a larval food plant. After addition of water and sequential extraction into ether, ethyl acetate, and *n*-butanol to remove compounds of increasing polarity, Nishida (1977) found that the aqueous layer remained very active and that little activity was present in any of the organic phases. Activity was not retained on a charcoal column, but paper chromatography of the active effluent failed to produce an active band that could be analyzed further.

### D. The Tribe Papilionini: Umbellifer-Feeding Species

The umbellifer-feeding species within the genus *Papilio* seem to represent a relatively recent stage in the adaptive radiation of the Papilionidae. Apparently derived from Rutaceae-feeding species, the approximately 11 species of the "*P. machaon* complex" are primarily temperate rather than tropical in distribution, and they feed as larvae on herbaceous umbellifers or on perennial herbs or shrubs of the genus *Artemisia* in the family Compositae (Asteraceae). Links with the Rutaceae are maintained in several species, including *P. zelicaon, P. polyxenes, P. brevicauda,* and *P. indra,* which will feed on both rutaceous and umbelliferous plants (Tyler, 1975). Umbellifer feeding has occasionally arisen elsewhere in the genus among species groups that are usually associated with the Rutaceae. The African species *P. demodocus,* for example, feeds on both umbellifers and rutaceous plants (van Son, 1949; Clarke *et al.*, 1963) and a South American species, *P. paeon,* attacks wild parsnip, *Pastinaca sativa* (Walker, 1882). In view of the close links between umbellifer-feeding swallowtails and some of the Rutaceae feeders, it would not be surprising to find similarities in the chemical cues used in host location, as suggested originally by Dethier (1941).

It appears that the only detailed investigation of chemical aspects of oviposition behavior among umbellifer-feeding swallowtails has been our study of the black swallowtail, *Papilio polyxenes*. Because none of this work has been published previously, the results will be summarized in some detail. In central New York, where the study is being conducted, the larvae feed primarily on the introduced umbellifers wild carrot (*Daucus carota*), wild parsnip (*Pastinaca sativa*), and poison hemlock (*Conium maculatum*) (Tietz, 1972; Shapiro, 1974). Before widespread deforestation during the past 200 years or so, the larvae presumably fed mainly on native marshland umbellifers such as *Sium suave, Cicuta bulbifera, Cicuta maculata,* and *Angelica atropurpurea* (see Tietz, 1972), and perhaps also on certain native woodland umbellifers such as *Osmorhiza* spp., *Cryptotaenia canadensis, Thaspium barbinode,* and *Taenidia integerrima* (see Tietz, 1972; Rehr, 1973; Scriber and Finke, 1978; Berenbaum, 1978).

After mating at hilltop lek sites (Lederhouse, 1982), *P. polyxenes* females fly across open fields, roadsides, or other habitats containing their host plants, often covering many hundreds of yards in a few minutes. At intervals during such flights, they drop down and lay a single egg on an umbelliferous plant before immediately resuming flight. Vision probably plays a major role in oviposition of this species and perhaps involves the sequential formation of search images for particular plants, as described by Rausher (1978) for *Battus philenor* females. The leaves of carrot and poison hemlock, generally preferred in laboratory choice tests, share a dissected, feathery appearance that might be used as a visual cue. Ovipositing females in the field have been seen to approach and apparently investigate plants of yarrow, *Achillea millefolium* (Asteraceae), which is not a larval host but which has leaves and flower heads superficially resembling those of carrot.

Once they have alighted upon a plant, ovipositing females of *Papilio polyxenes* engage in drumming behavior, usually followed immediately by abdomen-curling and oviposition if, as is usual, the plant belongs to a typical larval host species. Though no ablation or electrophysiological experiments have been conducted, there is every reason to believe that perception of host stimuli during drumming is mediated, as in other butterflies, by receptors located on the outer tarsal segments, especially of the forelegs (see Fig. 1). Like Nishida (1977) and Ichinosé and Honda (1978), the focus thus far has been on studying the contact-recognition phase of oviposition behavior. Specifically, attempts have been made to isolate and identify compounds in the foliage of carrot, *Daucus carota,* that stimulate abdomen-curling and egg laying by females of *P. polyxenes* once they have come into contact with a leaf surface. The working hypothesis has been that contact recognition is mediated primarily by one or more substances belonging to a key class of compounds that is relatively characteristic of apioid umbellifers, several rutaceous genera, and perhaps of other plant families attacked by swallowtails.

The butterflies used in the present research come from a year-round culture maintained in greenhouses and a controlled-environment chamber. Locally caught wild butterflies of both sexes are introduced to the culture during the summer months to minimize any effects of inbreeding. Larvae are raised primarily on the foliage of *Daucus carota* and *Pastinaca sativa* collected from the field during the summer months and grown in the greenhouse during the winter. Females are marked individually after emergence, hand-paired (Clarke and Sheppard, 1956) when 1 day old, and fed honey water (1:10) each morning. On days when they are needed for experiments, they are deprived of oviposition plants until required for bioassay in the afternoons. Because the rate of oviposition declines rapidly during the second week of life (Blau, 1981), females are generally not used for bioassay experiments after their tenth day of adult life.

Early experiments in this laboratory used a free-flight method, comparable to that used for *Colias* butterflies by Stanton (1979), to assay the responses of *P. polyxenes* females to extracts of carrot foliage in the laboratory. Females were introduced into large cages (1 × 1 × 1 m) covered with fine white gauze cloth and each containing an artificial "plant." The "plant" consisted of a wire stem, about 0.5 m tall, from the top of which several wire branches, about 15 cm long, extended outward horizontally in a symmetrical pattern. An artificial triangular leaf (6-cm long × 3.5-cm wide at the base × 0.8-cm thick), made from white or yellow domestic (ammonia-free) cellulose sponge (Robert F. Ogdon Co., Rochester, New York) and colored green by dipping in green food-coloring solution (McCormick and Co., Baltimore, Maryland), was pierced onto the end of each wire branch and secured with glue (Elmer's "Glue-all," Borden, Inc., Columbus, Ohio). For each experiment, the upper surface of each leaf was treated with 1 ml of test solution or solvent (control). The bioassay cages were placed beneath 60-W incandescent or infrared reflector lamps in a controlled-environment chamber (27.5°C, 60% RH, 1000 fc), and the butterflies allowed to lay eggs for up to 4 hr (Fig. 1c). The numbers of eggs on the experimental and control leaves were compared at the end of the experiment.

On the basis of the findings of Dethier (1941) and Saxena and Prabha (1975) that *Papilio* larvae are attracted to various essential oils, and on the widespread distribution of these compounds in the Umbelliferae, Rutaceae, and other plant families attacked by swallowtails (Table II), it was expected that the contact-oviposition response would be stimulated by steam distillates and by extracts of carrot leaves in nonpolar solvents. Early experiments by J. D. Hare in this laboratory showed that steam distillates of carrot leaves were ineffective as contact-oviposition stimulants (Table III). Ethanolic extracts of carrot leaves stimulated oviposition when applied to leaves on the artificial plants, but this activity was retained in the aqueous layer after removal of ethanol and extraction with ether (Table III). The active compounds thus appeared to be quite polar.

During these early experiments, it was noticed that pretreatment of sponge leaves with mineral oil (Squibb) to slow the rate of release of volatile compounds had the effect of enhancing the response of butterflies to aqueous fractions (Table III). For subsequent experiments, therefore, all leaves were routinely pretreated with 0.5 ml mineral oil. The free-flight design was further modified by standardizing the number of leaves per plant at four (opposite pairs of experimental and control leaves). Also, to control for any effects on the results of qualitative variation in ambient odor (see Vaidya 1969b), the procedure was adopted of mounting the artificial plants in the soil of pots containing live carrot plants. These were screened with green gauze cloth to prevent the butterflies from seeing them, while permitting the carrot odor to permeate the oviposition cages.

Use of the modified free-flight assay permitted satisfactory confirmation that compounds stimulatory to females are extracted into the aqueous phase of an ethanolic extract of carrot leaves, and that these compounds remain in the aqueous phase even after exhaustive extraction by chloroform, ethyl acetate, and *n*-butanol (Table IV). It is not as certain that the organic solvents did not remove some active components, however, because bioassay results were not always consistent (Table IV). Assays of Daucusöl (a commercial preparation of essential oils from carrot; Carl Roth GmbH, Karlsruhe, West Germany) and of methylisoeugenol, an essential oil component of carrot known to play a role in the oviposition behavior of the carrot rust fly *Psila rosae* (Berüter and Städler, 1971), proved negative (Table IV). The highly polar nature of the stimulant compounds confirms that these cannot be free essential oils, a conclusion reached also by Nishida (1977) for *Papilio xuthus* and by Ichinosé and Honda (1978) for *Papilio protenor*.

The free-flight bioassay method has the advantage of being about as close a

**TABLE III.** Oviposition by *Papilio polyxenes* Females on Arrays of Sponge Leaves Treated with Extracts or Steam Distillates of Carrot Leaves[a]

| Test Substrate[b,c] | Total Eggs Laid/Leaf | |
|---|---|---|
| | Cage 1 | Cage 2 |
| *Experiment A (5 females/cage)* | | |
| Water extract | 18 | 46 |
| Distilled water | 0 | 0 |
| Mineral oil | 0 | 1 |
| Mineral oil + water extract | 37 | 77 |
| *Experiment B (5 females/cage)* | | |
| Ethanol extract | 1 | 10 |
| Water extract | 16 | 12 |
| Distilled water | 1 | 0 |
| Water + steam distillate | 0 | 0 |
| Water extract + steam distillate | 9 | 2 |
| *Experiment C (11 females/cage)* | | |
| Water extract | 0 | 0 |
| Mineral oil + water extract | 37 | 66 |
| Mineral oil + water | 0 | 0 |
| Water + steam distillate | 0 | 0 |
| Water extract + steam distillate | 1 | 2 |
| Mineral oil + water + steam distillate | 0 | 0 |
| Mineral oil + water extract + steam distillate | 20 | 39 |

[a] "Free-flight" bioassay, 1976. Experiments conducted by J. D. Hare in authors' laboratory.

[b] Ethanol extract: 200 g of fresh carrot (*Daucus carota*) leaves were boiled in absolute ethanol (1800 ml) for 90 min, filtered, concentrated *in vacuo* to 200 ml, and refiltered. Water extract: 100 ml of ethanol extract were evaporated to dryness *in vacuo* at 50°C; the residue was extracted with diethyl ether and then dissolved in water. After washing three times with more ether, the aqueous layer was concentrated *in vacuo* at 50°C to 100 ml. Steam distillate: 300 g of fresh carrot leaves were ground in about 100 ml water, and steam was passed through the slurry for 90 min; the distillate was extracted three times with diethyl ether, and the combined ether fractions concentrated *in vacuo* to 5 ml.

[c] Substrates were applied to sponge leaves in the following order: Mineral oil (0.5 ml), water extract or (distilled) water (1 ml), steam distillate (1 ml).

simulation of a natural oviposition environment as we can expect to achieve easily in the laboratory. It has the disadvantage, however, of being slow. For further research the method was replaced by a filter-paper bioassay similar to the procedures described by Nishida (1977) and Ichinosé and Honda (1978). As before, female butterflies are fed in the morning and deprived of oviposition plants for 3–5 hr before bioassay in the controlled-environment chamber during the afternoon. Measured amounts (usually 100 μl) of test solutions or of a carrot

**TABLE IV.** Oviposition by *Papilio polyxenes* Females on Four-Leaved Artificial Plants Treated with Extracts of Carrot Leaves or with Essential Oils[a]

| | | | Total Eggs Laid per Leaf | | | |
|---|---|---|---|---|---|---|
| | | | Experimental | | Control[d] | |
| Date | Test Substrate[b,c] | Number of Females | A | B | A | B |
| (1977) | | | | | | |
| 7 July | Ethanol extract | 4 | 3 | 8 | 0 | 0 |
| 8 July | Ethanol extract | 4 | 10 | 8 | 0 | 0 |
| 9 July | Ethanol extract | 3 | 5 | 7 | 0 | 0 |
| 20 July | Water extract | 2 | 36 | 8 | 7 | 2 |
| 25 July | Water extract | 3 | 12 | 9 | 0 | 4 |
| 27 July | Water extract | 4 | 5 | 41 | 1 | 3 |
| 28 July | Water extract | 3 | 9 | 1 | 0 | 0 |
| 3 Aug. | Water extract | 4 | 4 | 56 | 0 | 0 |
| 9 Aug. | Water extract | 3 | 7 | 24 | 1 | 1 |
| 12 Aug. | Water extract | 3 | 52 | 3 | 0 | 0 |
| 17 Aug. | Water extract | 2 | 0 | 16 | 0 | 0 |
| 3 Sept. | Water extract | 3 | 14 | 37 | 7 | 3 |
| 8 Sept. | Water extract | 3 | 38 | 14 | 0 | 0 |
| 5 Oct. | Water extract | 3 | 39 | 51 | 0 | 1 |
| 13 July | Ethanol extract | 2 | 18 | 8 | 0 | 0 |
| | Ether extract | 2 | 0 | 0 | 0 | 0 |
| | Water extract | 2 | 6 | 7 | 0 | 0 |
| 14 July | Ether extract | 2 | 0 | 0 | 0 | 0 |
| | Water extract | 2 | 9 | 4 | 0 | 0 |
| 17 Oct. | Chloroform extract | 3 | 21 | 15 | 2 | 5 |
| | Residual water fraction | 3 | 0 | 0 | 0 | 0 |
| 18 Oct. | Water extract | 3 | 0 | 23 | 3 | 3 |
| | Chloroform extract | 3 | 0 | 0 | 0 | 0 |
| | Residual water fraction | 3 | 5 | 10 | 0 | 0 |
| 20 Oct. | Chloroform extract | 4 | 0 | 0 | 0 | 0 |
| | Residual water fraction | 4 | 13 | 2 | 0 | 1 |
| 20 Dec. | Water extract | 5 | 1 | 13 | 4 | 0 |
| 22 Dec. | *n*-Butanol extract | 5 | 7 | 11 | 1 | 0 |
| 23 Dec. | Water extract | 4 | 0 | 2 | 0 | 0 |
| | *n*-Butanol extract | 4 | 10 | 8 | 4 | 0 |
| (1978) | | | | | | |
| 3 Jan. | Chloroform extract | 5 | 0 | 0 | 0 | 0 |
| | *n*-Butanol extract | 5 | 0 | 4 | 0 | 0 |
| | Residual water fraction | 5 | 6 | 36 | 13 | 22 |
| 4 Jan. | Chloroform extract | 5 | 0 | 0 | 0 | 0 |
| | *n*-Butanol extract | 5 | 0 | 0 | 0 | 0 |
| | Residual water fraction | 5 | 84 | 127 | 3 | 1 |
| 5 Jan. | Chloroform extract | 5 | 0 | 0 | 0 | 0 |
| | *n*-Butanol extract | 5 | 0 | 0 | 0 | 0 |
| | Residual water fraction | 5 | 33 | 19 | 0 | 0 |

**TABLE IV.** *(Continued)*

| Date | Test Substrate[b,c] | Number of Females | Total Eggs Laid per Leaf | | | |
|---|---|---|---|---|---|---|
| | | | Experimental | | Control[d] | |
| | | | A | B | A | B |
| 11 July | Methylisoeugenol (1% in EtOH) | 2 | 0 | 0 | 0 | 0 |
| | Methylisoeugenol (0.1% in EtOH) | 2 | 0 | 0 | 0 | 0 |
| | Methylisoeugenol (0.01% in EtOH) | 2 | 0 | 0 | 0 | 0 |
| 12 July | Carrot oil (1% in EtOH) | 2 | 0 | 0 | 0 | 0 |
| | Carrot oil (0.1% in EtOH) | 2 | 0 | 0 | 0 | 0 |
| | Carrot oil (0.01% in EtOH) | 2 | 0 | 0 | 0 | 0 |

[a]Free-flight design, 1977–1978.
[b]Ethanol extract: Fresh carrot leaves (300–500 g) were ground in 80% ethanol (1–2 liters) in a blender and filtered. The extract was concentrated *in vacuo* to a volume equal to the original fresh weight of carrot tissue (300–500 ml). Water and ether extracts: Ethanol extract (200 ml) was concentrated *in vacuo* until all ethanol had been removed. To the remaining brown aqueous syrup was added about 50 ml water and 50 ml diethyl ether. After separation, the aqueous layer was extracted with fresh ether for 48 hr in a continuous liquid–liquid extractor before being made up to 200 ml (water extract). Combined ether washings and extracts were also made up to 200 ml (ether extract). Chloroform and *n*-butanol extracts: The water extract ($x$ ml) was extracted successively with chloroform, ethyl acetate, and *n*-butanol for 48 hr each in continuous liquid–liquid extractors. After each extraction, both the aqueous and organic phases were brought to the same volume as that of the starting water extract (i.e., $x$ ml).
[c]Mineral oil (0.5 ml) was first applied to all leaves, followed by 1 ml of distilled water (unless the test solution was aqueous), and finally by 1 ml of test solution.
[d]Control leaves were treated as experimentals except that 1 ml of the appropriate solvent was substituted for the test solution.

extract of known activity (''carrot-extract control'') are applied as bands, about 1 cm in width, toward the lower end of filter-paper strips (Whatman No. 1, 25 × 2 cm). Immediately before bioassay, each strip is moistened by misting with distilled water from an aerosol spray bottle. Each of several females is first placed by hand at the bottom of the carrot-extract control strip, which she walks up until encountering the band with her forelegs. Usually between 80 and 90% of females immediately curl their abdomens (Fig. 1d), a clear-cut response that was taken as a criterion for activity (whether or not it is followed by the laying of an egg). As during free-flight assays, a carrot plant is always present to provide odor during experiments.

Females giving a positive response to the carrot-extract control are next placed on a moistened blank strip, and any butterflies giving a positive response are rejected from further tests. The remaining females are placed in turn on the test strips and their responses recorded. Each female is given three opportunities to exhibit a positive response. After each test, every female giving a positive response is again placed on a distilled-water control; if she gives a positive response, she is rejected from further experiments that day. The overall activity of a test solution is scored as the fraction of unrejected females giving a positive response. Except for highly active extracts, reasonable reproducibility of results from one day to another cannot be expected unless at least 10 females are tested.

Individual females vary in their response thresholds to any given extract, the proportion of females giving a positive response increasing with concentration (Table V). On occasion, many of the females have become "hyperactive" after exposure to the carrot-extract control and have had to be rejected because of positive responses to distilled-water controls. In more recent bioassay experiments, therefore, a carrot plant rather than a carrot extract has been used for preliminary screening of females. A rigorous study of variability among females, including the role of phenotypic experience, must await identification of the stimulant compounds, after which it will be possible to establish concentrations precisely and control for any decomposition.

Though the filter-paper bioassay is used primarily to follow activity during further fractionation and purification of carrot extracts, a variety of pure compounds have also been tested in the hope of identifying attractant chemicals more quickly. As in the free-flight assay, carrot oil and methylisoeugenol proved negative (Table VI). Second only to essential oils, hydroxycoumarins and the related furanocoumarins are the most typical of the relatively unusual compounds linking the food plants of *Papilio* species (Table II); several of these compounds, especially in the form of glycosides, are polar enough to accord with the solubility properties that had already been established for the oviposition stimulants. However, none of the tested coumarins and furanocoumarins proved to be active (Table VI), and Berenbaum (1981b) failed to find evidence of furanocoumarins

**TABLE V.** Effect of Concentration on Responses by *Papilio polyxenes* Females to a Carrot Extract (Filter-Paper Bioassay)[a]

| Composition of Test Solution (μl)[b] | | | |
|---|---|---|---|
| Carrot Extract[c] | Distilled Water | Number of Females | Response |
| 6.25 | 93.75 | 14 | 0.43 |
| 12.50 | 87.50 | 13 | 0.62 |
| 25.00 | 75.00 | 13 | 0.69 |
| 50.00 | 50.00 | 12 | 0.75 |
| 100.00 | 0 | 11 | 0.82 |

[a]Fourteen females were selected 1 hr before the experiment, after giving positive responses to a carrot-extract control (100 μl) followed by negative responses to a distilled-water control (see text). Test solutions were presented to females in increasing order of extract concentration. A second experiment, in which extracts were presented in decreasing order of concentration, was marred by "hyperactivity" of most females following exposure to the most concentrated extract.

[b]All test and control strips were misted with water before bioassay.

[c]Aqueous layer from ethanol extract of fresh *Daucus carota* leaves, after solvent extraction with ether, chloroform, ethyl acetate, and *n*-butanol.

**TABLE VI.** Oviposition Responses of *Papilio polyxenes* Females to Pure Compounds (Filter-Paper Bioassay)

| Compound[a] | Response[b] | Compound[a] | Response[b] |
|---|---|---|---|
| Essential oils | | Flavonoids | |
| Methylisoeugenol | 0.00 (18) | Chrysin | 0.00 (25) |
| Geraniol | 0.13 (30) | Apigenin | 0.00 (25) |
| Carrot oil | 0.04 (26) | Apigenin 7-glucoside | 0.23 (13) |
| Coumarins and furanocoumarins | | | 0.00 (24) |
| Coumarin | 0.00 (25) | Apiin (apigenin 7-apiosylglucoside) | 0.00 (21) |
| Umbelliferone | 0.10 (19) | Kaempferol | 0.00 (23) |
| | 0.05 (18) | Morin | 0.00 (25) |
| Herniarin (7-methoxycoumarin) | 0.00 (18) | Myricetin | 0.00 (23) |
| Daphnetin (7,8-dihydroxycoumarin) | 0.00 (18) | Quercetin | 0.00 (25) |
| Skimmin (umbelliferone glucoside) | 0.09 (23) | Myricitrin (myricetin 3-rhamnoside) | 0.00 (23) |
| | 0.04 (24) | Quercitrin (quercetin 3-rhamnoside) | 0.00 (18) |
| Aesculin (aesculetin 6-glucoside) | 0.00 (25) | Hyperin (quercetin 3-galactoside) | 0.00 (23) |
| Xanthotoxin (8-methoxypsoralen) | 0.04 (23) | Hesperetin | 0.00 (23) |
| Imperatorin | 0.04 (23) | Naringenin | 0.00 (23) |
| Apterin | 0.00 (21) | Hesperidin (hesperetin 7-rutinoside) | 0.04 (26) |
| Hydroxycinnamic acids | | Naringin (naringenin 7-rhamnoside) | 0.00 (13) |
| *p*-Coumaric acid | 0.00 (25) | Fustin (dihydrofisetin) | 0.00 (26) |
| | 0.00 (21) | Furochromones | |
| Caffeic acid | 0.06 (16) | Khellin | 0.00 (25) |
| | 0.00 (18) | Aliphatic organic acids | |
| *o*-Hydroxycinnamic acid | 0.08 (24) | Citric acid | 0.00 (18) |
| *o*-Methoxycinnamic acid | 0.00 (24) | D(−)-Quinic acid | 0.05 (24) |
| Umbellic acid | 0.04 (25) | Lactic acid | 0.00 (18) |
| *o*-Glucosyloxycinnamic acid | 0.00 (16) | Malonic acid | 0.00 (18) |
| Chlorogenic acid | 0.28 (18) | | |
| | 0.21 (24) | | |
| | 0.24 (21) | | |

[a]Compounds were applied (100 μl) to test strips as concentrated (5–10%) solutions in an appropriate solvent (usually ethanol). Test strips were misted with water before bioassay.
[b]Fraction of acceptable females showing positive response. Number of females tested given in parentheses.

in carrot foliage. Organic acids are known to be accumulated in the tissues of rutaceous plants and of umbellifers, including *Daucus carota* (Robinson, 1975), but of the four compounds tested, none was active (Table VI).

It came to mind that many of the compounds characteristic of swallowtail food-plant families are derived biosynthetically from the phenylpropanoid pathway (Geissman and Crout, 1969). Such compounds include the coumarins and many of the essential oils. Although these compounds thus do not themselves seem to be responsible for activity, their correlation with swallowtail distribution could be accounted for if the active compounds are biosynthetic precursors or are related to such precursors. The central compounds in phenylpropanoid biosynthesis are the hydroxycinnamic acids, derived from phenylalanine (Fig. 4). Umbelliferone, the basic hydroxycoumarin and biosynthetic precursor of all the furanocoumarins, is formed simply from one of these hydroxycinnamic acids (Fig. 4) (Brown *et al.*, 1964; Austin and Meyers, 1965a, b). The most common hydroxycinnamic acids (caffeic, ferulic, *p*-coumaric, and *o*-coumaric) are widely distributed among plants and are not sufficiently polar to be likely candidates for our active material. Nevertheless we bioassayed some of them and indeed failed to find activity (Table VI). Intermediates in phenylpropanoid synthesis tend to occur as glucosides, which are considerably more polar than the parent compounds (Kosuge and Conn, 1961; Brown, 1962; Austin and Meyers, 1965b). We were able to obtain a sufficient quantity of only one such compound (*o*-glucosyloxycinnamic acid), however, and this proved to be inactive. Of all the pure compounds bioassayed, only chlorogenic acid gave a consistent positive response (Table VI).

Further purification of the active aqueous fraction from carrot leaves, following the series of extractions with organic solvents, has not proved simple or rapid. Activity could not be recovered reproducibly, if at all, from columns packed with silica gel, alumina, ion-exchange resins, or charcoal, apparently because the active compounds are easily degraded. The eventual choice was the

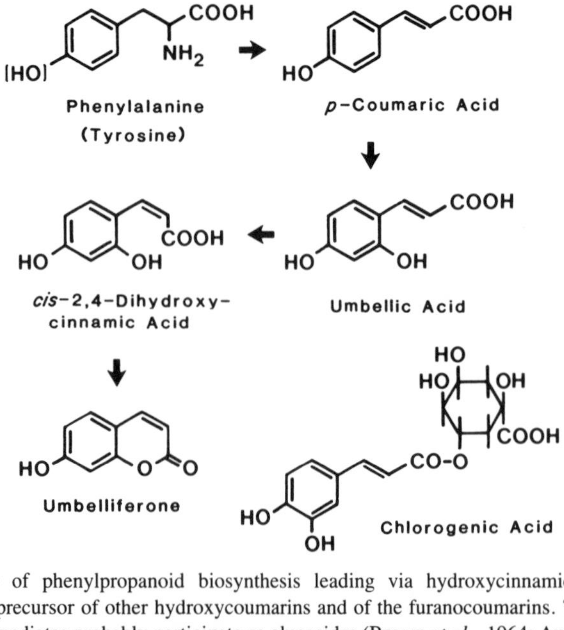

COOH
NH2
(HO)
Phenylalanine
(Tyrosine)
COOH
HO
*p*-Coumaric Acid
COOH
HO
OH
Umbellic Acid
COOH
HO
OH
*cis*-2,4-Dihydroxy-
cinnamic Acid
HO
O
O
Umbelliferone
HO
HO
OH
COOH
CO-O
HO
OH
Chlorogenic Acid

**Fig. 4.** Outline of phenylpropanoid biosynthesis leading via hydroxycinnamic acids to umbelliferone, the precursor of other hydroxycoumarins and of the furanocoumarins. The hydroxycinnamic acid intermediates probably participate as glucosides (Brown *et al.*, 1964; Austin and Meyers, 1965a,b). Chlorogenic acid is an ester of a hydroxycinnamic acid (caffeic acid) and quinic acid.

relatively mild Sephadex G-25 and LH-20 columns in sequence for preparative fractionation (Fig. 5). Activity was eluted in a single zone from the G-25 columns, coincident with the second brown band to elute (fraction F3, Fig. 5). Even on the Sephadex G-25 columns, however, attempts to follow activity quantitatively were frustrated by apparent decomposition. For example, the final water layer (165 ml) after solvent extraction of an extract of 3 kg (fresh weight) of domestic carrot leaves was found to contain at least 100 "butterfly units"/ml, one butterfly unit being the minimum quantity of material necessary to stimulate a positive response from at least 50% of 10 or more females in a filter-paper assay. After fractionation in batches on a large G-25 column and concentration of combined F3 fractions *in vacuo* to the original volume, activity had dropped to no more than 20 butterfly units/ml.

It was discovered that addition of the concentrated (aqueous) F3 fraction to a large excess of absolute ethanol produced a white precipitate, an aqueous solution of which was sometimes active to butterflies. The supernatant (always highly active) was concentrated *in vacuo,* applied to Sephadex LH-20 columns, and eluted with methanol or ethanol. Activity eluted from LH-20 columns is generally concentrated in fractions L4 (the second half of a colorless zone between two yellow flavonoid bands) and L5 (the second yellow band) (Fig. 5), though we have sometimes found activity in other fractions (see e.g., Table VII). As with the G-25 columns, eluted fractions were less active, after adjustment for concentration, than the starting material. This loss could be due to separation of two or more compounds the qualitative presence of which is required for optimum response. Recombination of fractions and suitable adjustment of concentration seemed to enhance activity in one experiment but had no effect in another. It seems most likely that a major source of loss of activity on LH-20 columns, as in other purification steps, is progressive degradation of active material.

The active L4 fraction from LH-20 columns was analyzed by high-pressure liquid chromatography (HPLC). Resolution of the highly polar components proved troublesome but limited success was achieved on a μBondapak $C_{18}$ analytical column (Waters Associates) eluted with water (reverse-phase). Peaks eluted in two regions of the chromatogram, corresponding to very polar compounds (VP) and polar compounds (P); other less polar compounds (undoubtedly including degradation products) were eluted as soon as the eluent was changed from water to ethanol. Bioassay of collected fractions revealed that activity was often associated with the VP region and, more specifically, with a peak we called Zone 2 (eluting after 8–9 min at a flow rate of 0.7 ml/min). Promising though this analytical result appeared to be, it proved impossible to apply on a preparative scale. Preparative columns do not give adequate resolution of peaks, and the repeated separations on analytical columns needed to accumulate adequate quantities of Zone 2 for bioassay and spectroscopic analysis take several weeks. With

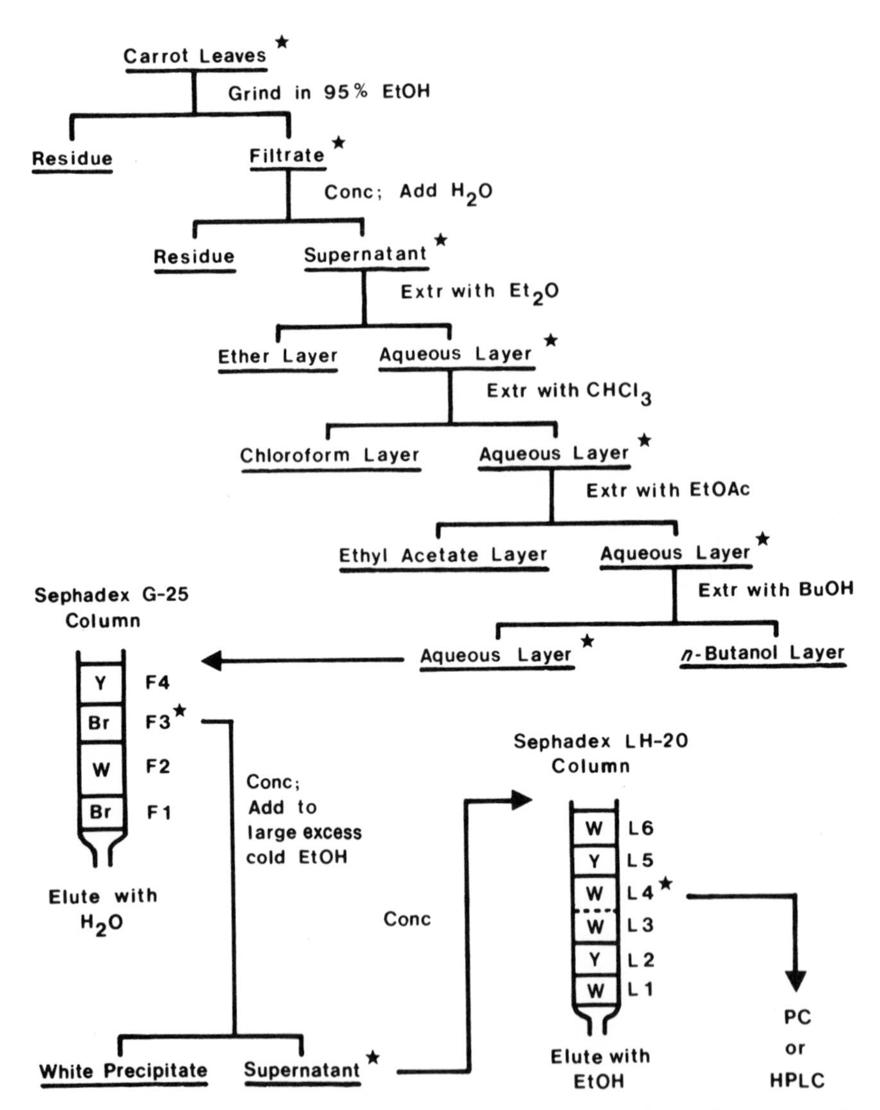


**Fig. 5.** Procedure used for fractionating ethanolic extracts of the leaves of carrot, *Daucus carota*. In recent extractions, the solvent extraction with ether is replaced by centrifugation to remove gummy, insoluble material. Fractions that are consistently active to female butterflies in oviposition bioassays are indicated by a star. Solvent extractions were conducted for 48 hr each in continuous liquid–liquid extractors. All concentration steps were carried out *in vacuo* below 50°C in a rotary evaporator. PC, Paper chromatography; HPLC, high-pressure liquid chromatography. Colors of zones eluted from Sephadex columns: Y, yellow; W, white (colorless); Br, brown.

**TABLE VII.** Effect of Hydrolysis of Active Fractions from Sephadex LH-20 Columns with Acid or β-Glucosidase on Stimulation of Oviposition by *Papilio polyxenes* Females (Filter-Paper Bioassay)

| Test Solution | Response[a] | |
|---|---|---|
| | Replicate 1 | Replicate 2 |
| Experiment 1[b] | | |
| LH-20 Fraction L4–L5 (control) | 1.00 (13) | 0.55 (11) |
| Fraction L4–L5 after acid hydrolysis | | |
| a. Ethyl acetate extract | 0 (13) | 0 (11) |
| b. Residual water layer | 0.08 (13) | 0 (11) |
| Fraction L4–L5 after enzyme hydrolysis | | |
| a. Ethyl acetate extract | 0 (13) | 0 (11) |
| b. Residual water layer | 1.00 (13) | 0.82 (11) |
| Experiment 2[c] | | |
| Fraction L2 | 0.30 (10) | — |
| Fraction L2 after acid hydrolysis | 0 (10) | — |
| Fraction L2 after enzyme hydrolysis | 0.30 (10) | — |
| Fraction L4–L5 | 1.00 (10) | — |
| Fraction L4–L5 after acid hydrolysis | 0 (10) | — |
| Fraction L4–L5 after enzyme hydrolysis | 0.60 (10) | — |

[a]Fraction of females, previously screened on control strips, giving a positive abdomen-curling response. The numbers of females tested are given in parentheses.
[b]One 130-μl aliquot of combined fractions L4 and L5 was incubated with 130 μl of 2 *N* HCl at 100°C for 2 hr and then neutralized with NaOH. A second aliquot was digested for 12 hr at 37°C with 1 mg β-glucosidase (Sigma Corp., St. Louis), dissolved in 130 μl of 0.5 *M* sodium acetate, adjusted to pH 5.5 with dilute acetic acid (Mabry *et al.*, 1970). A third aliquot (control) was maintained for 12 hr at 37°C after addition of 130 μl of the sodium acetate–acetic acid buffer (pH 5.5). Hydrolysates were extracted five times with ethyl acetate. Aqueous and ethyl acetate layers were concentrated *in vacuo* to the original 130 μl, which was applied to test strips for bioassay.
[c]Hydrolysis conditions were as in Experiment 1, except that aliquots were 100 μl.

the small amount of Zone 2 obtained by repeated peak-trapping, it was possible to establish that its ultraviolet spectrum displays a single broad peak at 260 nm, consistent with an aromatic structure. Separations by HPLC were hampered also by degradation of the active material as it was purified. Reinjection of accumulated Zone 2, collected as a single peak, revealed a mixture of peaks, including apparent regeneration of certain other components in both the P and VP regions of the chromatogram.

Hydrolysis of active LH-20 material (fractions L2 and L4–L5) with dilute mineral acid destroyed activity, but hydrolysis with β-glucosidase did not (Table VII). This suggests (but does not prove) that the active compounds are not simple glucosides. The LH-20 fractions could be separated into numerous components by paper chromatography (PC) in any of several solvent systems, including *n*-butanol–ethanol–water (4:1:2.2), isopropanol–water (4:1), and water (Smith, 1960; Harborne, 1967, 1973; Ribereau-Gayon, 1972). By allowing female butterflies to walk up the moistened chromatograms, it was possible to establish that active compounds had moved away from the origin, though it was impossible to correlate activity with any one spot on account of the complexity of the mixtures. Bioassay results were invariably better on Whatman No. 1 than on Whatman 3 MM papers (or on thin-layer plates). It was clear from the behavior of chromatogram spots in ultraviolet light that many of them, including several with $R_f$ values within zones of activity, were flavonoids (presumably present as glycosides or sulfated glycosides). The presence of flavonoid sulfates in fractions L2 and L4–L5 was consistent with "comet-shaped" spots on paper chromatograms and with the white precipitates formed by addition of aqueous $BaCl_2$ solution to both fractions after hydrolysis with 4 *N* HCl (Mabry *et al.*, 1970). Flavonoids had not been considered to be likely candidates for oviposition stimulants because they are generally distributed among angiosperms and show no obvious correlations with swallowtail food-plant distributions. Nevertheless, solutions of 16 flavonoids, chosen to represent various structural categories, were bioassayed with generally negative results (Table VI). A moderate response of 0.23 to apigenin 7-glucoside, one of the major flavonoid components of carrot leaves (Teubert and Herrmann, 1977), could not be repeated in a subsequent test (Table VI).

The components of active LH-20 fractions were separated somewhat successfully on columns of polyvinylpyrrolidone (Polyclar AT, General Aniline and Film Corp.), eluted with various proportions of water and methanol. On one occasion, elution of a 18 × 2.5-cm column with water yielded well-defined activity in the 50- to 90-ml fraction. In other separations, however, activity was either lost or spread weakly through several fractions. Pronounced color changes (to pinks, greens, and browns) accompanied Polyclar separations, suggesting extensive degradation. Paper electrophoresis (1000 V; 4–6 hr) also yielded erratic results. In some experiments, activity moved toward the anode in 0.2 *M* borate buffer at pH 8.2, whereas in other experiments activity was either lost or remained at the origin.

Because the active compounds are clearly similar in polarity to some of the flavonoid glycosides, the two-way PC system advocated by Mabry *et al.* (1970) seemed a worthwhile method for separating such compounds. After applying a sample to the origin on Whatman No. 1 or 3 MM papers (25 × 25 cm), the papers were run (ascending) first in *tert*-butanol–acetic acid–water (3:1:1 by volume) and then, after drying, in 15% acetic acid (aqueous). The LH-20 fractions gave rise to a reproducible pattern of spots, revealed by their fluorescence or absorbance in UV light before and after fuming with ammonia (Fig. 6). We adopted this system on a preparative scale, separating LH-20 fractions on batches of 20 Whatman 3 MM papers (57 × 46 cm, descending solvent flow), two tanks

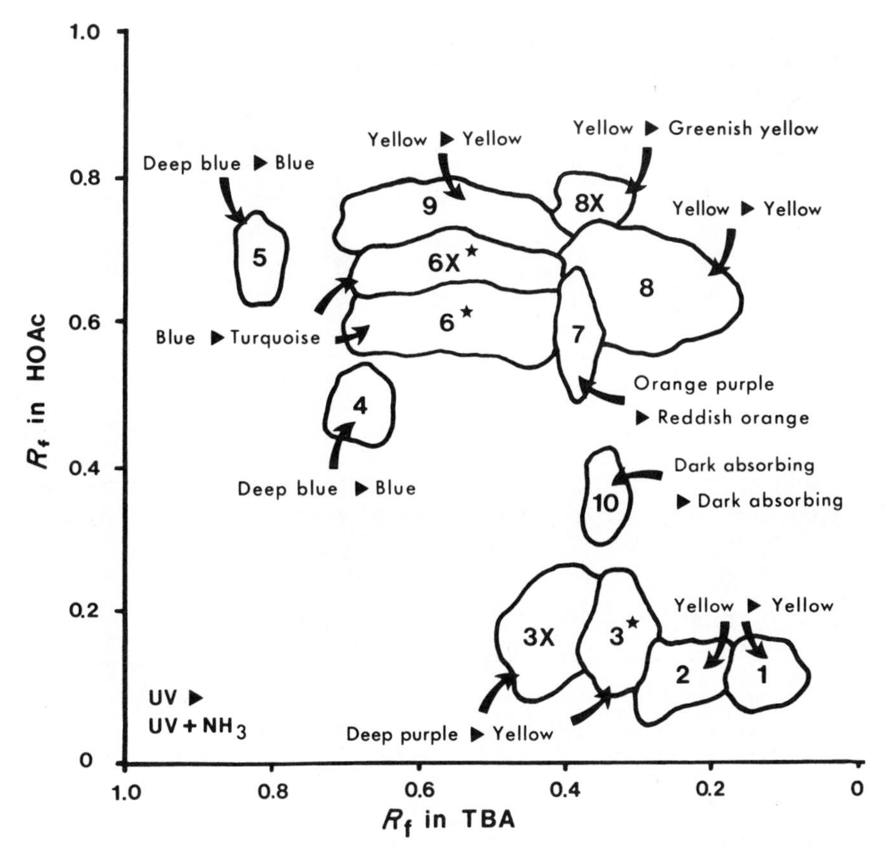


**Fig. 6.** Chromatographic pattern resulting from two-way paper chromatography of Sephadex LH-20 fractions, first in *tert*-butanol–acetic acid–water (3:1:1) (TBA), and second in 15% acetic acid (HOAc). Samples are applied to the origins of Whatman No. 1 or 3 MM papers and developed in ascending solvents. Zones are recognized by fluorescence colors in UV light (365 nm), before and after fuming with ammonia vapor (as indicated). Zones usually found to be active to butterflies in oviposition assays are indicated by stars.

at a time. From each of the papers the individual spots or zones were cut out and eluted with 50% aqueous ethanol or methanol. After concentration *in vacuo* these eluates were tested for activity using the filter-paper bioassay.

Three zones (3, 6, and 6X) were consistently active to butterflies (Table VIII). Zone 3, on ascending PC in water, yielded five UV fluorescent spots that also gave positive responses to ammoniacal $AgNO_3$ and to ferric chloride–potassium ferricyanide reagent (indicating a phenolic structure). The major spot ($R_f$ 0.28), corresponding to Compound 3 itself, was active to butterflies. Two minor spots ($R_f$ values 0.2 and 0.54) were apparently contaminants introduced when cutting

**TABLE VIII.** Oviposition Responses of *Papilio polyxenes* Females to Eluted Zones from Two-Dimensional Paper Chromatograms of Active Carrot Fractions (L4–L5) from LH-20 Sephadex Columns (Filter-Paper Bioassay)[a]

| Zone | Date (1981) | Number of Females | Response | Zone | Date (1981) | Number of Females | Response |
|---|---|---|---|---|---|---|---|
| 1 | 3/03 | 13 | 0.08 | 5 | 10/09 | 26 | 0.08 |
| | 3/12 | 23 | 0 | | 10/16 | 30 | 0.03 |
| | 10/16 | 30 | 0 | 6 | 3/12 | 23 | 0 |
| 2 | 3/03 | 13 | 0 | | 3/16 | 10 | 0.30 |
| | 3/12 | 23 | 0 | | 6/16 | 18 | 0.61 |
| | 3/16 | 10 | 0.20 | | 6/17 | 15 | 0.47 |
| 3 | 3/03 | 13 | 0.08 | | 6/18 | 16 | 0.81 |
| | 3/12 | 23 | 0.22 | | 8/10 | 21 | 0.14 |
| | 3/12 | 23 | 0.22 | | 10/07 | 25 | 0.08 |
| | 3/12 | 23 | 0.13 | | 10/19 | 30 | 0.13 |
| | 3/16 | 10 | 0.40 | 6X | 6/18 | 18 | 0.61 |
| | 4/15 | 23 | 0.43 | 7 | 3/12 | 23 | 0.04 |
| | 10/08 | 30 | 0.07 | | 3/16 | 10 | 0.10 |
| | 10/09 | 30 | 0.07 | 8 | 3/12 | 23 | 0 |
| 3X | 3/12 | 23 | 0 | | 10/14 | 30 | 0.17 |
| | 3/16 | 10 | 0.10 | | 10/19 | 30 | 0.03 |
| 4 | 3/12 | 23 | 0 | 8X | 3/12 | 23 | 0 |
| 5 | 3/12 | 23 | 0 | 10 | 10/14 | 30 | 0.03 |
| | 3/16 | 10 | 0.10 | | 10/14 | 28 | 0.14 |
| | 10/07 | 25 | 0.04 | | 10/16 | 30 | 0.07 |

[a]Batches of 10 large (57 × 46 cm) Whatman 3 MM papers at a time were run first in *t*-butanol–acetic acid–water (3:1:1) and then in 15% aqueous acetic acid. Chromatograms were examined under UV light to locate individual zones by their characteristic fluorescent colors. For bioassay, individual zones from 40 to 50 chromatograms were cut out and eluted with 50% aqueous ethanol. The eluates were concentrated *in vacuo* to 0.5 ml and 100-μl aliquots applied to test papers and misted with water before bioassay. Zone 9 has not yet been effectively assayed, due to accidental loss of eluate.

out Zone 3 from two-way chromatograms and were inactive. The remaining two spots ($R_f$ values 0.67 and 0.77) corresponded in properties and $R_f$ values on rechromatography in TBA–acetic acid with those of Compounds 6 or 6X (blue fluorescent) and 8 or 8X (yellow fluorescent). The spot corresponding to 6 or 6X was the more intense of the two and was active to butterflies.

The overlapping Zones 6 and 6X (eluted together from the two-way chromatograms) yielded four or five overlapping spots at $R_f$ 0.79–0.85 when chromatographed on paper in water. Rechromatography of Zones 6–6X in the two-way system (TBA–acetic acid) revealed two major spots, corresponding to Compounds 6 and 6X themselves, accompanied by smaller amounts of compounds corresponding in properties to the neighboring Zones 4, 7, and 8 or 8X (see Fig. 6). There was no evidence for any formation of Compound 3 from 6–6X, analogous to the apparent formation of 6 or 6X on rechromatography of 3.

The properties of Compounds 3, 6, and 6X, as revealed by UV light and spraying with various diagnostic reagents, are reminiscent of those of hydroxycinnamic acid derivatives (Table IX). The quenching of UV fluorescence by Benedict's reagent indicates the presence of two unsubstituted hydroxyl groups located ortho to one another on an aromatic nucleus, as, for example, in caffeic and chlorogenic acids (Reznik and Egger, 1961) (Fig. 4). Such dihydroxyphenols are notoriously susceptible to oxidative degradation (Sondheimer, 1964), and a structure of this nature could account for the loss of activity that we have consistently encountered while purifying extracts, and for the apparent oxidative browning we have encountered when applying such extracts to paper chromatograms.

We selected the Zones 6 and 6X for further analysis; though not readily separated on paper chromatograms, they appeared to be more active to butterflies than was Zone 3 (Table IX) and also seemed to be more stable. We subjected the combined Zones 6–6X to acid hydrolysis (2*N* HCl or $H_2SO_4$ for 2 hr at 60°C) and extracted the hydrolysates with ethyl acetate. Paper chromatography of the ethyl acetate fraction in water revealed four major and four minor components (Table IX). One of the major components (Spot 6.7) was revealed only by its pink color with diazotized *p*-nitroaniline–$Na_2CO_3$ reagent but seemed to be the source of a sweet odor that had been noticed previously when active extracts were subjected to acid hydrolysis. Its nature remains unknown. Two of the other major components (Spots 6.2 and 6.5) corresponded in responses to UV light and to spray reagents with the two isomers of caffeic acid (Table IX); their lower $R_f$ values could have resulted from interference by other components of the hydrolysate mixture. The aqueous fractions of acid (HCl or trifluoroacetic acid) hydrolysates failed to yield a positive response for sulfate ions with barium chloride solution, and an initial search for sugars by thin-layer chromatography was likewise fruitless.

Mass spectroscopy has recently confirmed the presence of caffeic and quinic acids in ethyl acetate–diethyl ether (50:50) extracts of acid hydrolysates of Zone 6. After removal of the solvent in a stream of nitrogen, the components were converted to their trimethylsilyl (TMS) ethers. The resulting mixture contained five major and several minor components, which were resolved by gas chromatography (GC) on a 5-ft OV 101 column and then examined by GC mass spectroscopy. Two of the peaks corresponded exactly in retention time and mass spectrum to those of the TMS ethers of pure samples of caffeic and quinic acids, respectively (P. Feeny and L. Rosenberry, unpublished results). Based on a comparison of $R_f$ values in several solvent systems, alongside chlorogenic acid standards, it now seems likely that Zones 6 and 6X contain the trans and cis isomers of chlorogenic acid, respectively (M. B. Cohen and M. Carter, unpublished results).

The probable presence of chlorogenic acid in the active Zones 6–6X and the

**TABLE IX.** Paper Chromatography of Active Zones 3, 6, and 6X Eluted from Two-Way Chromatograms (TBA–15% Acetic Acid) and of Ethyl Acetate-Soluble Aglycones Released by Hydrolysis of Zones 6–6X[a]

| | $R_f$ Value (× 100)[b] | | | Responses to Light and Spray Reagents[c] | | | | | | |
|---|---|---|---|---|---|---|---|---|---|---|
| | | | | | | Diazotized *p*-nitroaniline | | Benedict's Reagent | | |
| Zone or Compound | TBA | HOAc (15%) | Water | UV (365 nm) | UV–$NH_3$ | Alone | $Na_2CO_3$ | Daylight | UV | Sodium Nitrite |
| Unhydrolyzed active zones | | | | | | | | | | |
| Zone 3 | 31–34 | 16–21 | 21–28 | Purple[d] | Yellow | Orange | Brown | Yellow | Quen | Yel-bn |
| Zone 6 | 55–67 | 65–74 | 79–83 | Blue | Turq | Orange | Brown | Yellow | Quen | Yel-bn |
| Zone 6X | 56–67 | 76–82 | 81–85 | Blue | Turq | Orange | Brown | Yellow | Quen | -ive |
| Hydrolysis products of zones 6–6X | | | | | | | | | | |
| Spot 6.1 | — | — | 4–5 | Yellow | Yellow | Pink | Gry-bn | (Faint) | Quen | — |
| Spot 6.2[e] | — | — | 21–26 | Blue | Bt blue | Gry-bn | Gry-bn | Yellow | Quen | — |
| Spot 6.3 | — | — | 29 | (Faint) | Gn-yel | -ive | Brown | (Faint) | (Faint) | — |
| Spot 6.4 | — | — | 36–62 | (Faint) | Pur-bl | Orange | Brown | (Faint) | (Faint) | — |
| Spot 6.5[e] | — | — | 48–67 | Blue | Turq | Orange | Brown | Yellow | Quen | — |
| Spot 6.6[e] | — | — | 66–71 | Yellow | Yellow | Orange | Brown | Yellow | Gn-yel | — |
| Spot 6.7[e] | — | — | 77–80 | -ive[f] | -ive[f] | -ive | Pink | -ive | -ive | — |
| Spot 6.8 | — | — | 83–86 | Abs | Abs | -ive | -ive | -ive | -ive | — |

| | | | | | | | | | | |
|---|---|---|---|---|---|---|---|---|---|---|
| Reference compounds | | | | | | | | | | |
| Chlorogenic acid | — | — | 73–89 | Blue | Turq | Orange | Brown | Yellow | Quen | Yellow |
| Caffeic acid | — | — | 71–77 | Blue | Bt blue | Orange | Gry-bn | Yel-bn | Quen | Yellow |
| | | | 24 | | | | | | | |
| *o*-Coumaric acid | — | — | 86,80 | Blue | Gn-yel | Yellow | Pink | Gn-yel | Gn-yel | — |
| *p*-Coumaric acid | — | — | 80–86 | (Faint) | Blue | Orange | Blue | — | — | — |
| Ferulic acid | — | — | 78 | Purple | Blue | Pink | Blue | (Faint) | Blue | — |
| Umbelliferone | — | — | 48–59 | Blue | Bt blue | Yel-or | Pink-bn | Yellow | Blue | — |
| Scopoletin | — | — | 35 | Blue | Bt blue | Gry-bl | Blue | Yellow | Blue | — |
| Umbellic acid | — | — | 80,47 | Blue | Bt blue | Orange | Gry-bn | Pink | Blue | — |
| Apterin | — | — | 75 | Purple | Purple | — | — | -ive | -ive | — |

[a]Hydrolysis conditions as in Table VII, footnote *b*.

[b]Solvent systems: TBA, *t*-butanol–acetic acid–water (3:1:1); 15% HOAc, 15% acetic acid in water. $R_f$ Values were quite variable, depending on temperature, complexity and concentration of the sample, and nature of the paper (Whatman No. 1 or 3 MM).

[c]Spray reagents prepared as follows. For diazotized *p*-nitroaniline, $NaNO_2$ (5 drops of a 5% solution in $H_2O$) and sodium acetate (8 ml of a 20% aqueous solution) were added to *p*-nitroaniline (2 ml of a 0.5% solution in 2 *N* HCl). Colors were recorded before and after spraying with 15% aqueous $Na_2CO_3$ (Swain, 1953). For Benedict's reagent, $CuSO_4 \cdot 5H_2O$ (1.73 g), sodium citrate (17.3 g), and anhydrous $Na_2CO_3$ (10 g) were dissolved in water and made up to 100 ml (Stahl, 1969). Sodium nitrite, 2% solution in 50% aqueous ethanol (Block *et al.*, 1958). Bt blue, bright blue; Gry-bn, gray-brown; Gn-yel, greenish-yellow; Turq, turquoise; Abs, dark, UV-absorbing spot; Gry-bl, grayish-blue; Yel-or, yellowish-orange; Pink-bn, pinkish-brown; Pur-bl, purplish-blue; Quen, color quenched; Yel-bn, yellowish-brown; -ive, no response; —, not tested.

[d]Appears orange-brown by reflected light, purple by transmitted light.

[e]Major components.

[f]After heating (105°C) in an oven for a few minutes, the central area only of this spot fluoresced yellow in UV, turned darker yellow on fuming with $NH_3$, and thereafter appeared pale brown in daylight. Spot 6.7 turned yellow when sprayed with bromcresol green solution (Sigma Corp., St. Louis), indicating the presence of an acid.

consistently appreciable response of butterflies to solutions of pure chlorogenic acid suggest that this compound may be a natural oviposition stimulant for *Papilio polyxenes*. It is still possible, however, that the active compounds in Zones 6–6X are related molecules, such as other isomers of caffeoylquinic acid or quinic acid esters of hydroxycinnamic acids to which other molecules are attached by second ester or glycosidic linkages (Sondheimer, 1964). Production of the sweet odor on hydrolysis suggests the presence of a terpenoid or phenylpropanoid essential oil component in bound form, perhaps as a hydroxycinnamic acid derivative. Chlorogenic acid itself is unlikely to be responsible by itself for the contact response in nature; it is too widely distributed and its activity never reaches that of our more active extracts. It could, however, represent part of an overall chemical profile to which the butterflies respond. Chlorogenic acid is known to play such a role in host selection by the Colorado potato beetle,

**TABLE X.** Oviposition Responses of *Papilio polyxenes* Females to Washed Carrot Leaves and Leaf Washings, and to Extracts of Lemon Epicarp (Filter-Paper Bioassay)[a]

| Test Solution | Number of Females | Response |
|---|---|---|
| Experiment 1[b] | | |
| Distilled water | 28 | 0 |
| Carrot leaves | 28 | 1.00 |
| Washed carrot leaves | 28 | 0.92 |
| Experiment 2[b] | | |
| Distilled water | 15 | 0 |
| LH-20 fractions L4–L5 | 15 | 1.00 |
| Washed carrot leaves | 15 | 1.00 |
| Concentrated leaf washings | 10 | 0 |
| Experiment 3[c] | | |
| Extract of lemon epicarp (500 μl) | 19 | 0.47 |
| | 16 | 0.56 |
| | 18 | 0.22 |
| Extract of lemon epicarp after extraction | 18 | 0.22 |
| with organic solvents (100 μl) | 16 | 0.31 |

[a]Experiments 1 and 2 by Phillip S. Bose in authors' laboratory.
[b]Bunches of freshly cut carrot leaves (~ 50 g) were immersed, except for the cut ends of the stems, in 700 ml distilled water and agitated gently for 5 min (Exp. 1) or 30 min (Exp. 2). They were rinsed in water and shaken gently to remove excess water before bioassay. Control carrot leaves were simply held upside down in air for the same periods. Leaf washings (Exp. 2) were concentrated *in vacuo* at 45°C to 0.3 ml and applied to a paper for bioassay.
[c]The epicarp tissue (yellow outer portion) of eight lemons (*Citrus limon*) was blended in 95% ethanol and filtered. After the residue had been rinsed twice with boiling ethanol, the combined filtrates were concentrated to 10 ml *in vacuo* (see Ichinosé and Honda, 1978); 5 ml of this concentrate were extracted three times each with ethyl acetate and *n*-butanol.

*Leptinotarsa decemlineata* (Hsiao and Fraenkel, 1968; Mitchell and Schoonhoven, 1974).

The stimulant compounds, whatever their identity, appear to be located inside the tissues of carrot leaves and not on the leaf surfaces. Leaves soaked in distilled water for up to 30 min and assayed immediately remained fully active to butterflies, whereas the concentrated leaf washings were inactive (Table X). Leaves heated for 30 min in 2 *N* HCl at 40°C remained fully active. Only under rather severe conditions, accompanied by obvious signs of tissue disruption, did leaves lose their activity (Table XI). Because activity is known to be acid-labile *in vitro* (Table VII), these results suggest that one or more of the required components is located within the leaf cuticle.

## E. Interspecific Comparisons of Oviposition Responses to Plant Extracts

If a certain compound, or any one of a closely related group of compounds, is sufficient to stimulate oviposition behavior by swallowtails, as would be predicted by the sign stimulus hypothesis in its simplest form, one might expect females of related species to respond similarly to the same plant extracts. Nishida (1977) indeed found that females of *Papilio machaon hippocrates,* an umbellifer-feeding species, responded positively to methanolic extracts of two rutaceous plants, *Poncirus trifoliata* and *Zanthoxylum piperitum,* that are larval

**TABLE XI.** Oviposition Responses of *Papilio polyxenes* Females to Carrot Leaves Exposed to Hydrochloric Acid Solutions for 30 Min at Different Temperatures[a]

| | Temperature (°C) | | | | |
|---|---|---|---|---|---|
| Acid Strength (*N*) | 30 | 40 | 60 | 80 | 100 |
| 0 | 1.00 | 1.00 | 1.00 | 0.53[c] | 0[c] |
| 0.1 | 1.00 | 1.00 | 0.92 | 0.53[b] | 0[b,c] |
| 0.25 | 1.00 | 1.00 | 0.75 | 0.33[b] | 0[b,c] |
| 0.5 | 1.00 | 1.00 | 0.58[b] | 0.27[b] | 0[b,c] |
| 1.0 | 1.00 | 1.00 | 0.33[b] | 0[b,c] | 0[b,c] |
| 2.0 | 1.00[b] | 1.00[b] | 0.33[b] | 0[b,c] | 0[b,c] |

[a]Experiment conducted by Phillip Bose at authors' laboratory. A different live carrot plant was used for each temperature. Fronds from the plant were bent over into six beakers, placed on electric hotplates arranged radially around the plant. Each beaker contained distilled water or one of the five HCl solutions, in which the leaves were submerged. After 30 min, leaves were cut from the plant, rinsed in distilled water, and tested for activity to 15 females. Bioassay procedure was that used for filter-paper assays (see text).
[b]Leaves had turned completely brown after 30 min.
[c]Acid solution had become distinctly colored after 30 min.

food plants of *Papilio xuthus*. Similarly, females of *Papilio polyxenes* respond positively to ethanolic extracts of lemon fruit (*Citrus limon*) (Table X) and *Zanthoxylum americanum* (R. Hagen, unpublished results). However, Nishida (1977) found that females of *P. xuthus* and *P. macilentus,* another Rutaceae-feeding species, would not respond to extracts of *D. carota,* a food plant of *P. machaon*.

More recently, Abe *et al*. (1981) have found that responses of three Japanese Rutaceae-feeding *Papilio* species to ethanolic extracts of each others' food plants are by no means identical. Females of *P. bianor,* for example, are strongly stimulated (fraction of females responding = 0.70) by the extract of their own host species, *Orixa japonica,* whereas the same extract elicited very little response (0.07) from females of *P. xuthus* and *P. protenor*. Females of *P. protenor* were similarly disinterested (0.03) in an extract of *Phellodendron amurense* that was highly stimulatory to *P. xuthus* (0.80) and *P. bianor* (0.83). By contrast, extracts of *Citrus limon, Poncirus trifoliata,* and *Fagara ailanthoides* were stimulatory to females of all three species.

Reciprocal relationships between different swallowtail genera seem to be even less close. Nishida (1977) found that females of *Byasa (Parides) alcinous* and *Luedhorfia japonica* would respond positively to methanolic extracts only of their own host plants, *Aristolochia debilis* and *Asarum takaoi,* respectively. Females of neither species responded to extracts of *D. carota* (Umbelliferae) or of *Poncirus trifoliata* or *Zanthoxylum piperitum* (Rutaceae). It remains to be seen whether different genera of swallowtails respond to different classes of compounds as key stimulants or whether the same stimulant compounds are used but combined with responses to different arrays of other chemical ingredients.

## V. DISCUSSION

The sequence of behaviors leading up to oviposition by butterflies can be divided into three phases—namely, habitat search, plant search, and plant assessment—though this sequence is interrupted by other behaviors such as nectar-feeding and basking. During the first two phases of oviposition behavior, females fly to an appropriate habitat and search within it for potentially suitable larval food plants, which they approach. During the third phase, a plant is assessed for the degree to which its chemical profile falls within some acceptable range (usually including those profiles characteristic of larval food species) and for qualities such as water content (Wolfson, 1980), size, and age (Jones, 1977; Ives, 1978), that may be correlated with larval growth success. The probability of subsequent oviposition may also be affected by the presence of other eggs or larvae (Rothschild and Schoonhoven, 1977; Rausher, 1979; Shapiro, 1980), or of structures that mimic insect eggs (Williams and Gilbert, 1981). In practice it is not always possible to distinguish successive phases of oviposition behavior from one another. The act of landing on a plant, for example, would seem to mark a convenient boundary between plant search and plant assessment, but there appears to be considerable variation among different species in the extent to which females can assess the identity and suitability of plants before alighting on them.

Vision appears to play the dominant role during the first two phases of oviposition behavior, both for assessing gross habitat variables, such as light intensity, and for guiding insects to plants of particular colors or shapes (see Papaj and Rausher, Chapter 3, this volume). The role of olfaction during habitat and plant search remains unclear. Despite the early confidence expressed by Verschaffelt (1911) and Brues (1920), orientation of females to host odors seems to have been demonstrated only for *Papilio demoleus* (Saxena and Goyal, 1978), as discussed earlier. There is also circumstantial evidence to suggest that searching females of *Pieris brassicae* are attracted to allylnitrile and perhaps other volatile compounds emanating from wild cabbage, *Brassica oleracea* (Mitchell, 1977). In most other butterfly species investigated, however, visual responses seem sufficient to account for the observed behavior. Laboratory experiments with *Papilio xuthus* (Nishida, 1977), *Papilio protenor* (Ichinosé and Honda, 1978), *Papilio polyxenes* (this chapter), and *Pieris rapae* (J. A. Renwick, personal communication) have so far failed to show consistent orientation toward volatile components of appropriate food plants. There are, however, several examples of volatile compounds acting as repellents or oviposition inhibitors (e.g., Hovanitz and Chang, 1964; Lundgren, 1975; Rothschild and Schoonhoven, 1977).

Whether watching butterflies in the field or conducting preference tests in the laboratory, it is generally impossible to know when host odors are completely lacking. If females orient to visual cues only in the presence of certain odor components, as reported by Vaidya (1969a) for *Papilio demoleus,* and if the threshold concentrations of these components are low, it might be exceedingly difficult to rule out odor as an important cue for searching females. Similar difficulties arise in assessing the role of host odors in orientation from longer distances or in triggering anemotaxis by flying females (see Schütte, 1966). Faced with an analogous problem, however, S. Finch and G. Skinner (see Feeny, 1982) were able to demonstrate convincingly that anemotaxis by female cabbage root flies, *Delia radicum,* is not dependent on host odors; these are detected only at a range of a few meters from the host plant. It remains to be seen how general this pattern might be.

In contrast to the role of plant chemistry during earlier phases of oviposition behavior, the importance of chemical stimuli during the plant-assessment phase of oviposition behavior is general and well established. Observations and experiments with species from all families of butterflies have demonstrated beyond doubt that stimuli perceived by tarsal chemoreceptors during drumming behavior are of paramount importance. Whether or not receptors located elsewhere also play a role remains in doubt and may vary widely among species. Chemotactile hairs have been found on the ovipositors of *Pieris brassicae* females but their function remains unknown (Klijnstra, 1982). Antennectomized females of *P. brassicae, Danaus gilippus,* and *Papilio protenor* appear to lay eggs normally (Myers, 1969; Ma and Schoonhoven, 1973; Ichinosé and Honda, 1978), but the curious antennal dipping behavior that accompanies drumming by *C. lacinia* females (Calvert, 1974) suggests that the antennae may aid plant assessment in some species. The tarsal receptors are known to respond to solutions and thus to serve as contact chemoreceptors. Whether or not they can also respond to odors is apparently not known (see Städler, 1977).

Of considerable interest is the nature of the compounds that are perceived by tarsal chemoreceptors and that stimulate oviposition itself. Experiments with several species suggest that the presence of water may be universally necessary, a finding consistent with the identification of a water receptor as one of the five cells in the B hairs of *Pieris brassicae* (Ma and Schoonhoven, 1973). That two of the other four cells in B hairs are salt receptors suggests that the presence of salts may also be important, though preliminary experiments with *Papilio polyxenes* show no synergistic effect of sodium chloride solutions on the ovipositional response to moistened chlorogenic acid (P. Feeny, L. Rosenberry, and M. Carter, unpublished results). That one of the remaining cells in the B hairs of *P. brassicae* responds to solutions of glucosinolates (Ma and Schoonhoven, 1973), and that females of this species readily lay eggs on green paper treated only with allylglucosinolate (David and Gardiner, 1962) have provided compelling reasons for believing that chemicals belonging to a single and relatively unique class of secondary compounds play a decisive role as key compounds or sign stimuli in host assessment by butterflies generally. There is even circumstantial evidence to suggest that differences in oviposition preference for related plants may sometimes result simply from the abilities of females to discriminate among related compounds belonging to such a single class (Rodman and Chew, 1980).

As found for lepidopterous larvae and in other insect orders, however, reality may prove more complex, and the apparent simplicity of the relationship between pierids and glucosinolates may be misleading (Dethier, 1976). Among papilionids, both the present investigators and Abe *et al.* (1981) are finding that the stimulant activity of host-plant extracts divides into several fractions on purification, each usually of lower activity than the parent extract. Recombination of fractions sometimes leads to enhanced activity (Abe *et al.*, 1981). The patterns are such as to suggest that we are not dealing merely with different compounds of the same general class but, rather, with compounds of two or more classes. Even among pierids, there is reason to doubt that glucosinolates are the whole answer. J. A. Renwick (personal communication) has found that several fractions of extracts from cruciferous plants, including a relatively nonpolar fraction, have a stimulating effect on the contact-oviposition response by *Pieris rapae*.

Contact receptors may respond to compounds that inhibit rather than stimulate oviposition. Egg laying by pierid females, for example, can be reduced by treating host leaves with extracts of various nonhost species (Lundgren, 1975; Rothschild and Fairbairn, 1980). Contact chemoreception is partly responsible for inhibition of oviposition by *Pieris brassicae* females in the presence of conspecific eggs or larvae (Behan and Schoonhoven, 1978).

From studies on chemoreception in lepidopterous larvae, Dethier (1976, 1982) has concluded that receptor cells responding to vapors or solutions are mostly generalists, each responding to a variety of different compounds and providing a complex pattern of input to the central nervous system where behavioral "decisions" are made. Although a particular compound or class of compounds may contribute the major sensory cue, it is the total chemical complex that forms the basis for perception. Such a broader view, the "response spectrum" hypothesis, seems entirely in accord with the pattern that seems to be emerging from studies of oviposition behavior in butterflies. It could perhaps be broadened yet further to include integration of chemical input with that from mechanoreceptors and visual stimuli.

Laboratory investigators are faced with a perplexing and often frustrating degree of variability in the oviposition behavior of experimental butterflies, which contrasts oddly with the typical predictability and precision of oviposition by butterflies in the field. Some of this variability must presumably be due directly to genetic differences among individuals, but there is good reason to believe that much and perhaps most of it results from environmental variation and phenotypic experience (Chew and Robbins, in press). Schoonhoven (1976) has found that the impulse patterns generated by the taste sensilla of caterpillars, when stimulated by a given stimulus (e.g., a plant sap), may vary with the age of the individual insect and also with previous experience (in this case, diet). Even though he could not find evidence for interactive effects among the different cells within a sensillum, Schoonhoven (1976) emphasizes that independent variation of the responses of different cells as a result of age and experience can lead by itself to a bewildering variety of input patterns to the central nervous system. Rausher (1978) and Traynier (1979) have demonstrated that prior experience by ovipositing females of *Battus philenor* and *Pieris rapae*, respectively, can markedly alter subsequent responses to visual and chemical cues. Females of *Euphydryas editha* become progressively less discriminating in their preference for oviposition sites as the time of unsuccessful search increases (Singer, 1971, 1982). Variation in behavior can also be expected to result from other differences in motivation of individuals, depending on physiological state, from even subtle variations in the relative concentrations of chemical constituents, and from interactions between chemical and other stimuli (see Dethier, 1982). Understanding the nature of variability among individual butterflies is a major challenge, not made any easier by incomplete qualitative knowledge of the cues used in oviposition behavior.

Ehrlich and Raven (1964) offered a hypothesis to explain the correlations they had documented between the food-plant relationships of butterflies and patterns of secondary chemistry in plants. Following Fraenkel (1959), they proposed that many of the characteristic natural products in plants have been evolved, and presumably still are being evolved, as a consequence of the selective pressures exerted on plants by phytophagous insects and other herbivores. Butterflies have occasionally evolved adaptations permitting them to overcome the barriers in novel food plants. Thereafter, facilitated by preadaptation at the toxicological level, these insects were able to colonize new "adaptive zones" (Ehrlich and Raven, 1964). Once established within such a zone, further evolutionary divergence could presumably follow from long-term association with the novel plant taxon and its chemically similar descendents (Chew and Robbins, in press). The stepwise evolution of resistance by plants and adaptations for tolerance by enemies, seen as a "dynamic equilibrium" by Dethier (1954), was christened "coevolution" by Ehrlich and Raven (1964; see also Janzen, 1980), though essentially the same process had been referred to earlier as "reciprocal adaptive evolution" (Fraenkel, 1959) and as "parallel evolution" (Brues, 1920). Jermy (1976), deemphasizing the role of herbivores in selecting for plant compounds, referred to shifts in the host relationships of phytophagous insects as "sequential evolution."

Unlike Brues (1920) and Dethier (1941), Ehrlich and Raven (1964) apparently considered that behavioral changes followed rather than preceded physiological changes during colonization events: "After the restriction of certain groups of insects to a narrow range of food plants, the formerly repellent substances of these plants might, for the insects in question, become chemical attractants [p. 602]." It is hard to imagine, however, how a toxicological breakthrough by butterfly larvae can result in a permanent colonization event unless it is associated with behavioral mechanisms for relocating the novel food plant in the next generation (Stride and Straatman, 1962). As put more recently by Dethier (1970), "the first barrier to be overcome in the insect–plant relationship is a behavioral one. The insect must sense and discriminate before nutritional and toxic factors become operative [p. 98]."

Although some host shifts have probably been initiated by changes in larval behavior, it is simpler to visualize the evolution of feeding preferences in most butterfly groups as having resulted from changes in the oviposition behavior of female adults. Females may occasionally lay eggs on novel food plants as a result of sudden and rather radical mutations in sensory ability ("maternal instinct"), as foreseen by Brues (1920) and discussed in some detail by Dethier (1970). Probably more frequently, and accompanied by little or no mutation, females lay eggs on novel plants that share with the normally preferred plants one or more of the (usually chemical) characteristics used as behavioral cues in host location, as suggested (for larvae) by Dethier (1941). The probability of oviposition on novel or less preferred hosts would be increased by changes in ecological conditions such as a decrease in density of the preferred host species (Singer, 1971). However initiated, the subsequent fate of the attempted colonization will be influenced by larval success. In addition to surviving predation and other sources of mortality associated with the novel plant, the larvae must tolerate any nutritional abnormalities and toxins in the new food. Any adults that result are likely to inherit not only the ability of larval progeny to tolerate toxins in the novel plant but also, perhaps, some genetically based behavioral predisposition to choose the novel plant for oviposition. Depending on ecological conditions such as the relative abundances of the novel and original food-plant species (Singer, 1971), and on the butterfly's population size and its mating system (Gilbert, 1979), the colonization may or may not then succeed as a relatively permanent phenomenon with selection acting to reinforce preference for the novel host (Singer, 1971). Patterns of chemical similarity among the food plants of related butterflies may thus result in part from differences in the rates of successful colonization. Attempted colonizations are more frequent on plants of similar chemistry, because of preadaptation at the behavioral level, and a greater proportion of attempted colonizations are successful on plants of similar chemistry, owing to preadaptation at the toxicological level.

The foregoing scenario is consistent with both the sign stimulus and response spectrum hypotheses of host-plant recognition, though the latter hypothesis more readily explains some of the fine tuning observed in host-plant relationships. Of the various *Rutaceae*-feeding swallowtails studied in Japan by Abe *et al*. (1981), for example, some will not oviposit on certain of the food plants of others. This would be a surprising observation if all species shared just a single class of compounds serving as a sign stimulus, and yet it would be equally surprising to find that such closely related species have evolved responses to radically different classes of compounds. It seems much more likely that the butterflies have diverged in the overall chemical profiles to which they respond, though these profiles presumably retain major elements in common. Divergence in response spectra could occur merely as a consequence of prolonged reproductive isolation, or it might occur quite rapidly as a consequence of unique selective pressures acting on the different populations. Chew (1977) has suggested that one consequence of high mortality of *Pieris* larvae on plants of the genus *Thlaspi* (Cruciferae) might be selection for females that avoid laying eggs on plants of this genus. The average response spectrum of such females would presumably come to differ from that of females in other populations not exposed to *Thlaspi* plants, even though glucosinolates could remain as important components of the response spectra of all females. The patterns of host preference among crucifer-feeding flea beetles display clearly the contrast between universal feeding stimulation by glucosinolates and inhibitory responses of individual species to cucurbitacins and cardenolides (Nielsen, 1978).

It should be clear from the foregoing discussion not only that oviposition behavior is an important aspect of the present-day ecology of butterflies and their food plants, but also that it may have played a central role in the evolution of patterns of interaction between these two groups of organisms, as foreseen by Brues (1920). Because butterflies are relatively easy to observe in the field and to manipulate in the laboratory, and because their relationships with food plants are paralleled in many ways by those of other groups of phytophagous insects, they can serve as valuable models for testing ideas of more general applicability. It should also be apparent that the oviposition behavior of butterflies has long been thought to involve plant chemistry, and that there exists considerable circumstantial evidence to this effect. Only relatively recently, however, has direct experimental evidence for the role of chemistry been obtained, and we still know remarkably little about the compounds involved, the relative roles of taste and olfaction, the nature and degree of variability among females, and the role of phenotypic experience. Even in this comparatively well-studied group of insects there is clearly much to be learned.

## VI. SUMMARY

The proximate mechanisms by which female butterflies select oviposition sites largely determine patterns of host use in ecological time, with important consequences to the ecology of both the butterflies and the host plants. In evolutionary time, these mechanisms are thought to have influenced and even initiated longer term shifts in patterns of host-plant use by butterflies. In spite of the importance of these behavioral mechanisms, they have received relatively little attention until recently and in no case are the details of the behavioral sequences fully established.

The role of chemistry in oviposition behavior, suspected since early in this century, has been investigated by two approaches. Correlations between phytochemical patterns and the food-plant associations of related butterflies have provided circumstantial evidence that particular classes of secondary compounds play a key role in host-plant recognition; in at least one case, such deductions have received direct experimental support. Concurrently, investigations of chemoreceptors have shown that butterflies are capable of detecting and discriminating between a great variety of compounds, some acting as stimulants and some as inhibitors, some being secondary compounds and others not. Most attention has been devoted to contact chemoreception; the role of olfaction in oviposition behavior remains unclear.

Studies of contact chemoreception in several swallowtail species (family Papilionidae) indicate that the most active stimulants are polar compounds. No single class of key compounds, analogous to glucosinolates in pierid food plants, has yet been identified. Responses to such compounds, if they exist, are probably modified by the synergistic and antagonistic effects of several other plant components. Chlorogenic acid, for example, may play a subsidiary role in contact stimulation of oviposition by females of the black swallowtail, *Papilio polyxenes,* on carrot, *Daucus carota.*

Present evidence suggests that egg-laying females are capable of assessing several plant characteristics, including complex chemical profiles. Responses to such profiles may or may not be dominated by responses to one or more key ingredients. The sign stimulus hypothesis may be a special case of a more general response spectrum hypothesis. Responses of females to plant compounds are rendered more difficult to interpret by considerable individual variation, at least part of which seems to result from phenotypic experience.

## Acknowledgments

For help with field work or laboratory experiments, we sincerely thank R. V. Barbehenn, P. S. Bose, P. L. Chao, M. B. Cohen, Y. M. Eisner, J. D. Hare, L. Hunter, M. Juenger, A. Miller, J. S. Miller, J. A. Renwick, and A. L. Schrager. For the loan of equipment or gifts of compounds or column materials, we thank D. J. Austin, S. A. Brown, E. E. Conn, L. L. Creasy, K. Herrmann, T. Kosuge, T. J. Mabry, E. Städler, W. Steck, M. D. Whalen, and D. B. Wilson. For encouragement and advice, we are particularly indebted to Dr. May Berenbaum. The manuscript has benefited greatly from helpful comments by F. S. Chew, A. J. Damman, V. G. Dethier, R. Hagen, J. S. Miller, J. A. Renwick, K. N. Saxena, and E. Städler. We thank E. R. Hoebeke for taking the SEM photographs.

## REFERENCES

Abe, H., Teramoto, Y., and Ichinosé, T. (1981). Relationship between host plant ranges of the three papilionid butterflies and oviposition-inducing contact chemicals in their host plants. *Appl. Entomol. Zool.* **16,** 493–496.

Anderson, A. L. (1932). The sensitivity of the legs of common butterflies to sugars. *J. Exp. Zool.* **63,** 235–259.

Austin, D. J., and Meyers, M. B. (1965a). The formation of 7-oxygenated coumarins in Hydrangea and Lavender. *Phytochemistry* **4,** 245–254.

Austin, D. J., and Meyers, M. B. (1965b). Studies on glucoside intermediates in umbelliferone biosynthesis. *Phytochemistry* **4,** 255–262.

Behan, M., and Schoonhoven, L. M. (1978). Chemoreception of an oviposition deterrent associated with eggs in *Pieris brassicae. Entomol. Exp. Appl.* **24,** 163–179.

Berenbaum, M. (1978). *Taenidia integerrima,* a new foodplant record for *Papilio polyxenes. J. Lepid. Soc.* **32,** 303–304.

Berenbaum, M. (1980). ''Furanocoumarin Chemistry, Insect Herbivory, and Coevolution in the Umbelliferae.'' Ph.D. Dissertation, Cornell Univ., Ithaca, New York.

Berenbaum, M. (1981a). An oviposition ''mistake'' by *Papilio glaucus* (Papilionidae). *J. Lepid. Soc.* **35,** 75.

Berenbaum, M. (1981b). Patterns of furanocoumarin distribution and insect herbivory in the Umbelliferae: Plant chemistry and community structure. *Ecology* **62,** 1254–1266.

Berüter, J., and Städler, E. (1971). An oviposition stimulant for the carrot rust fly from carrot leaves. *Z. Naturforsch.* **26b,** 339–340.

Blau, W. S. (1981). Life history variation in the black swallowtail butterfly. *Oecologia* **48,** 116–122.

Block, R. J., Durrum, E. L., and Zweig, G. (1958). "A Manual of Paper Chromatography and Paper Electrophoresis," 2nd ed. Academic Press, New York.

Brown, S. A. (1962). Biosynthesis of the coumarins. III. The role of glycosides in the formation of coumarin by *Hierochloe odorata. Can. J. Biochem. Physiol.* **40,** 607–618.

Brown, S. A., Towers, G. H. N., and Chen, D. (1964). Biosynthesis of coumarins. V. Pathways of umbelliferone formation in *Hydrangea macrophylla. Phytochemistry* **3,** 469–476.

Brues, C. T. (1920). The selection of food-plants by insects, with special reference to lepidopterous larvae. *Am. Nat.* **54,** 313–332.

Calvert, W. H. (1974). The external morphology of foretarsal receptors involved with host discrimination by the nymphalid butterfly, *Chlosyne lacinia. Ann. Entomol. Soc. Am.* **67,** 853–856.

Chew, F. S. (1977). Coevolution of pierid butterflies and their cruciferous foodplants. II. The distribution of eggs on potential foodplants. *Evolution* **31,** 568–579.

Chew, F. S., and Robbins, R. K. (1983). Egg-laying in butterflies. *Symp. R. Entomol. Soc. London* **11,** in press.

Clarke, C. A., and Sheppard, P. M. (1956). Hand pairing of butterflies. *Lepid. News* **10,** 47–53.

Clarke, C. A., Dickson, C. G. C., and Sheppard, P. M. (1963). Larval color pattern in *Papilio demodocus. Evolution* **17,** 130–137.

Cronquist, A. (1968). "The Evolution and Classification of Flowering Plants." Houghton, Boston, Massachusetts.

David, W. A. L., and Gardiner, B. O. C. (1962). Oviposition and the hatching of the eggs of *Pieris brassicae* (L.) in a laboratory culture. *Bull. Entomol. Res.* **53,** 91–109.

Dethier, V. G. (1941). Chemical factors determining the choice of food plants by *Papilio* larvae. *Am. Nat.* **75,** 61–73.

Dethier, V. G. (1954). Evolution of feeding preferences in phytophagous insects. *Evolution* **8,** 33–54.

Dethier, V. G. (1970). Chemical interactions between plants and insects. *In* "Chemical Ecology" (E. Sondheimer, and J. B. Simeone, eds.), pp. 83–102. Academic Press, London and New York.

Dethier, V. G. (1976). The importance of stimulus patterns for host plant recognition and acceptance. *In* "The Host Plant in Relation to Insect Behavior and Reproduction" (T. Jermy, ed.), pp. 67–70. Plenum, New York.

Dethier, V. G. (1982). Mechanism of host-plant recognition. *Entomol. Exp. Appl.* **31,** 49–56.

Ehrlich, P. R., and Raven, P. H. (1964). Butterflies and plants: A study in coevolution. *Evolution* **18,** 586–608.

Eltringham, H. (1933). On the tarsal sense organs of Lepidoptera. *Trans. Entomol. Soc. London* **81,** 33–36.

Feeny, P. P. (1982). Ecological aspects of insect-plant relationships—round-table discussion. *Proc. 5th Int. Symp. Insect–Plant Relationships,* pp. 275–283. Pudoc, Wageningen, the Netherlands.

Fox, R. M. (1966). Forelegs of butterflies. I. Introduction: Chemoreception. *J. Res. Lepid.* **5,** 1–12.

Fraenkel, G. S. (1959). The *raison d'être* of secondary plant substances. *Science (Washington, D.C.)* **129,** 1466–1470.

Frings, H., and Frings, M. (1949). The loci of contact chemoreceptors in insects. *Am. Midl. Nat.* **41,** 602–658.

Frings, H., and Frings, M. (1956). The loci of contact chemoreceptors involved in feeding reactions of certain Lepidoptera. *Biol. Bull. (Woods Hole, Mass.)* **110,** 291–299.

Geissman, T. A., and Crout, D. H. G. (1969). "Organic Chemistry of Secondary Plant Metabolism." Freeman, San Francisco, California.

Gilbert, L. E. (1979). Development of theory in the analysis of insect–plant interactions. *In* "Analysis of Ecological Systems" (D. J. Horn, R. Mitchell, and G. R. Stairs, eds.), pp. 117–154. Ohio State Univ. Press, Columbus.

Gilbert, L. E., and Singer, M. C. (1975). Butterfly ecology. *Annu. Rev. Ecol. Syst.* **6,** 365–397.

Grabowski, C. T., and Dethier, V. G. (1954). The structure of tarsal chemoreceptors of the blowfly *Phormia regina* Meigen. *J. Morphol.* **94,** 1–19.

Guenther, E. (1948–1952). "The Essential Oils," Vols. 1–4. Van Nostrand, New York.

Harborne, J. B. (1967). Chromatography of phenolic compounds. *In* "Chromatography" (E. Heftmann, ed.), 2nd ed., pp. 677–698. Van Nostrand–Reinhold, New York.

Harborne, J. B. (1973). "Phytochemical Methods." Halsted, New York.

Hegnauer, R. (1964–1973). "Chemotaxonomie der Pflanzen," Vols. 3–6. Birkhäuser, Basel and Stuttgart.

Heywood, V. H. (ed.)(1971). "The Biology and Chemistry of the Umbelliferae." Academic Press, London and New York.

Hovanitz, W., and Chang, V. C. S. (1964). Adult oviposition responses in *Pieris rapae*. *J. Res. Lepid.* **3,** 159–172.

Hsiao, T. H., and Fraenkel, G. (1968). Isolation of phagostimulative substances from the host plant of the Colorado potato beetle. *Ann. Entomol. Soc. Am.* **61,** 476–484.

Ichinosé, T., and Honda, H. (1978). Ovipositional behavior of *Papilio protenor demetrius* Cramer and the factors involved in its host plants. *Appl. Entomol. Zool.* **13,** 103–114.

Ilse, D. (1937a). Aus dem Leben der Schmetterlinge (Film presentation). *Mitt. Dtsch. Entomol. Ges.* **7,** 84–88.

Ilse, D. (1937b). New observations on responses to colours in egg-laying butterflies. *Nature (London)* **140,** 544–545.

Ilse, D. (1956). Behaviour of butterflies before oviposition. *J. Bombay Nat. Hist. Soc.* **53,** 486–488.

Ilse, D., and Vaidya, V. G. (1956). Spontaneous feeding response to colours in *Papilio demoleus* L. *Proc. Indian Acad. Sci. Sect. B* **43,** 23–31.

Ives, P. M. (1978). How discriminating are cabbage butterflies? *Aust. J. Ecol.* **3,** 261–276.

Janzen, D. H. (1980). When is it coevolution? *Evolution* **34,** 611–612.

Jermy, T. (1976). Insect–host–plant relationship: Co-evolution or sequential evolution? *In* "The Host-Plant in Relation to Insect Behaviour and Reproduction" (T. Jermy, ed.), pp. 109–113. Plenum, New York and London.

Jones, R. E. (1977). Movement patterns and egg distribution in cabbage butterflies. *J. Anim. Ecol.* **46,** 195–212.

Klijnstra, J. W. (1982). Perception of the oviposition deterrent pheromone in *Pieris brassicae*. *Proc. 5th. Int. Symp. Insect–Plant Relationships,* pp. 145–151. Pudoc, Wageningen, the Netherlands.

Kosuge, T., and Conn, E. E. (1961). The metabolism of aromatic compounds in higher plants. III. The β-glucosides of *o*-coumaric, coumarinic and melilotic acids. *J. Biol. Chem.* **236,** 1617–1621.

Lederhouse, R. C. (1982). Territorial defense and lek behavior of the black swallowtail, *Papilio polyxenes*. *Behav. Ecol. Sociobiol.* **10,** 109–118.

Lundgren, L. (1975). Natural plant chemicals acting as oviposition deterrents on cabbage butterflies [*Pieris brassicae* (L.), *P. rapae* (L.) and *P. napi* (L.)]. *Zool. Scr.* **4,** 253–258.

Ma, W. C., and Schoonhoven, L. M. (1973). Tarsal contact chemosensory hairs of the large white butterfly *Pieris brassicae* and their possible role in oviposition behaviour. *Entomol. Exp. Appl.* **16,** 343–357.

Mabry, T. J., Markham, K. R., and Thomas, M. B. (1970). "The Systematic Identification of Flavonoids." Springer–Verlag, New York and Berlin.

Minnich, D. E. (1921). An experimental study of the tarsal chemoreceptors of two nymphalid butterflies. *J. Exp. Zool.* **33,** 173–203.

Minnich, D. E. (1922a). The chemical sensitivity of the tarsi of the red admiral butterfly, *Pyrameis atalanta* L. *J. Exp. Zool.* **35,** 57–81.

Minnich, D. E. (1922b). A quantitative study of tarsal sensitivity to solutions of saccharose in the red admiral butterfly, *Pyrameis atalanta* L. *J. Exp. Zool.* **36,** 445–457.

Mitchell, B. K., and Schoonhoven, L. M. (1974). Taste receptors in Colorado beetle larvae. *J. Insect Physiol.* **20,** 1787–1793.

Mitchell, N. D. (1977). Differential host selection by *Pieris brassicae* (the large white butterfly) on *Brassica oleracea* subsp. *oleracea* (the wild cabbage). *Entomol. Exp. Appl.* **22,** 208–219.

Morita, H., and Takeda, K. (1957). The electrical resistance of the tarsal chemosensory hairs of the butterfly *Vanessa indica. J. Fac. Sci. Hokkaido Univ. Series 6* **13,** 465–469.

Morita, H., and Takeda. K. (1959). Initiation of spike potentials in contact chemosensory hairs of *Vanessa. J. Cell. Comp. Physiol.* **54,** 177–187.

Morita, H., Doira, S., Takeda, K., and Kuwabara, M. (1957). Electrical response of contact chemoreceptors on the tarsus of the butterfly *Vanessa indica. Mem. Fac. Sci. Kyushu Univ. Ser. E* **2,** 119–139.

Munroe, E. (1960). The generic classification of the Papilionidae. *Can. Entomol.* (Suppl. 17), 1–51.

Munroe, E., and Ehrlich, P. R. (1960). Harmonization of concepts of higher classification of the Papilionidae. *J. Lepid. Soc.* **14,** 169–175.

Myers, J. (1969). Distribution of foodplant chemoreceptors on the female Florida queen butterfly, *Danaus gilippus berenice* (Nymphalidae). *J. Lepid. Soc.* **23,** 196–198.

Nielsen, J. K. (1978). Host plant discrimination within Cruciferae: Feeding responses of four leaf beetles (Coleoptera: Chrysomelidae) to glucosinolates, cucurbitacins and cardenolides. *Entomol. Exp. Appl.* **24,** 41–54.

Nishida, R. (1977). Oviposition stimulants of some papilionid butterflies contained in their host plants. *Botyu-Kagaku* **42,** 133–140.

Rausher, M. D. (1978). Search image for leaf shape in a butterfly. *Science* (*Washington, D.C.*) **200,** 1071–1073.

Rausher, M. D. (1979). Egg recognition: Its advantages to a butterfly. *Anim. Behav.* **27,** 1034–1040.

Rehr, S. S. (1973). New foodplant records for *Papilio polyxenes* F. (Papilionidae). *J. Lepid. Soc.* **27,** 237–238.

Reznik, H., and Egger, K. (1961). Benedicts Reagens als Indicator für phenolische *ortho*-Dihydroxygruppen. *Fresenius Z. Anal. Chem.* **183,** 196–199.

Ribereau–Gayon, P. (1972). "Plant Phenolics." Oliver & Boyd, Edinburgh.

Robinson, T. (1975). "The Organic Constituents of Higher Plants," 3rd ed. Cordus, North Amherst, Massachusetts.

Rodman, J. E., and Chew, F. S. (1980). Phytochemical correlates of herbivory in a community of native and naturalized Cruciferae. *Biochem. Syst. Ecol.* **8,** 43–50.

Rothschild, M., and Fairbairn, J. W. (1980). Ovipositing butterfly (*Pieris brassicae* L.) distinguishes between aqueous extracts of two strains of *Cannabis sativa* L. and THC and CBD. *Nature* (*London*) **286,** 56–59.

Rothschild, M., and Schoonhoven, L. M. (1977). Assessment of egg load by *Pieris brassicae* (Lepidoptera: Pieridae). *Nature* (*London*) **266,** 352–355.

Saxena, K. N., and Goyal, S. (1978). Host-plant relations of the citrus butterfly *Papilio demoleus* L.: Orientational and ovipositional responses. *Entomol. Exp. Appl.* **24,** 1–10.

Saxena, K. N., and Prabha, S. (1975). Relationship between the olfactory sensilla of *Papilio demoleus* L. larvae and their orientation responses to different odours. *J. Entomol. Ser. A* **50,** 119–126.

Schoonhoven, L. M. (1976). On the variability of chemosensory information. *Symp. Biol. Hung.* **16,** 261–266.

Schütte, von F. (1966). Beobachtungen zum Zug von Faltern der Gattung *Pieris* Schrk. *Z. Angew. Entomol.* **58,** 131–138.

Scriber, J. M. (1973). Latitudinal gradients in larval feeding specialization of the world Papilionidae (Lepidoptera). *Psyche* **80,** 355–373.

Scriber, J. M., and Finke, M. (1978). New foodplant and oviposition records for the eastern black swallowtail, *Papilio polyxenes,* on an introduced and a native umbellifer. *J. Lepid. Soc.* **32,** 236–238.

Seigler, D. S. (1977). Plant systematics and alkaloids. *In* "The Alkaloids" (R. H. Manske, ed.), Vol. 16, pp. 1–82. Academic Press, New York.

Shapiro, A. M. (1974). Butterflies and skippers of New York State. *Search.* **4,** 1–60.

Shapiro, A. M. (1980). Egg-load assessment and carryover diapause in *Anthocaris* (Pieridae). *J. Lepid. Soc.* **34,** 307–315.

Singer, M. C. (1971). Evolution of food-plant preference in the butterfly *Euphydryas editha. Evolution* **25,** 383–389.

Singer, M. C. (1982). Quantification of host specificity by manipulation of oviposition behavior in the butterfly *Euphydryas editha. Oecologia,* **52,** 224–229.

Smith, I. (ed.)(1960). "Chromatographic and Electrophoretic Techniques," Vol. 1. Wiley (Interscience), New York.

Sondheimer, E. (1964). Chlorogenic acids and related depsides. *Bot. Rev.* **30,** 667–712.

Städler, E. (1977). Sensory aspects of insect plant interactions. *Proc. Int. Congr. Entomol. (Washington, D.C., 1976),* pp. 228–248.

Stahl, E. (ed.)(1969). "Thin-Layer Chromatography: A Laboratory Handbook," 2nd ed. Springer–Verlag, Berlin and New York.

Stanton, M. L. (1979). The role of chemotactile stimuli in the oviposition preference of *Colias* butterflies. *Oecologia* **39,** 79–91.

Stride, G. O., and Straatman, R. (1962). The host plant relationship of an Australian swallowtail, *Papilio aegeus,* and its significance in the evolution of host plant selection. *Proc. Linn. Soc. N.S.W.* **87,** 69–78.

Swain, T. (1953). The identification of coumarins and related compounds by filter-paper chromatography. *Biochem. J.* **53,** 200–208.

Takeda, K. (1961). The nature of impulses of single tarsal chemoreceptors in the butterfly *Vanessa indica. J. Cell. Comp. Physiol.* **58,** 233–245.

Teubert, H., and Herrmann, K. (1977). Flavonols and flavones of vegetables. VIII. Flavones of carrot leaves. *Lebensm. Unters. Forsch.* **165,** 147–150.

Tietz, H. M. (1972). "An Index to the Described Life Histories, Early Stages and Hosts of the Macrolepidoptera of the Continental United States and Canada," 2 vols. A. C. Allyn, Sarasota, Florida.

Traynier, R. M. M. (1979). Long term changes in the oviposition behavior of the cabbage butterfly, *Pieris rapae,* induced by contact with plants. *Physiol. Entomol.* **4,** 87–96.

Tyler, H. A. (1975). "The Swallowtail Butterflies of North America." Naturegraph, Healdsburg, California.

Vaidya, V. G. (1956). On the phenomenon of drumming in egg-laying female butterflies. *J. Bombay Nat. Hist. Soc.* **54,** 216–217.

Vaidya, V. G. (1969a). Form perception in *Papilio demoleus* L. (Papilionidae, Lepidoptera). *Behaviour* **33,** 212–221.

Vaidya, V. G. (1969b). Investigations on the role of visual stimuli in the egg laying and resting behaviour of *Papilio demoleus* L. (Papilionidae, Lepidoptera). *Anim. Behav.* **17,** 350–356.

van Son, G. (1949). "The Butterflies of Southern Africa, Part I: Papilionidae and Pieridae," Memoir No. 3. Transvaal Museum, Pretoria.

Verschaffelt, E. (1911). The cause determining the selection of food in some herbivorous insects. *Proc. Acad. Sci. Amsterdam* **13,** 536–542.

Walker, J. J. (1882). A life history of *Papilio paeon* Roger. *Entomol. Mon. Mag.* **19,** 53–55.

Weis, I. (1930). Versuche über die Geschmacksrezeption durch die Tarsen des Admirals, *Pyrameis atalanta* L. *Z. Vgl. Physiol.* **12,** 206–248.

Williams, K. S., and Gilbert, L. E. (1981). Insects as selective agents on plant vegetative morphology: Egg mimicry reduces egg laying by butterflies. *Science (Washington, D.C.)* **212,** 467–469.

Wolfson, J. L. (1980). Oviposition response of *Pieris rapae* to environmentally induced variation in *Brassica nigra*. *Entomol. Exp. Appl.* **27,** 223–232.

# 3

# Individual Variation in Host Location by Phytophagous Insects*

DANIEL R. PAPAJ AND MARK D. RAUSHER



## I. INTRODUCTION

> Furthermore, accurate information regarding individual variation in host preferences among insects and the variability within polyphagous populations is completely lacking. Careful studies of feeding habit variability among populations of monophagous, oligophagous, and polyphagous species of insects would be invaluable.
>
> V. G. Dethier, 1954

*This work was supported by grants DEB 80-16414 and DEB 81-10218 from the National Science Foundation, and grants from the Duke University Research Council.



HERBIVOROUS INSECTS

ISBN 0-12-045580-3

A major objective of the study of how phytophagous insects locate and select host plants is an understanding of the behavioral processes that contribute to the limited range of host-plant types utilized by any given herbivore population. By understanding these mechanisms, we may then begin to comprehend how such behavior may be altered, both by natural selection and by humans. The very premise that host-seeking behavior may be altered, however, implies that such behavior is variable: different individuals search for host plants in different ways. In fact, evidence has begun to accumulate suggesting that variation between and within populations in host-selection behavior is a common phenomenon (Singer, 1971, 1983; Rausher, 1978, 1980; Fox and Morrow, 1981; Tabashnik *et al.*, 1981; Rausher and Papaj, in press). Moreover, we are gradually recognizing that variation in host selection often contributes to variation in patterns of host use between and within herbivore populations (Singer, 1971, 1982a; Chew, 1975, 1977; Wiklund, 1975; Rausher, 1980).

Despite this evidence, little is known about the mechanisms underlying individual differences in behavior. Nevertheless, only by assessing individual differences in behavior and determining the fundamental processes responsible for behavioral variation can we begin to perceive ways in which variability has been molded by natural selection to generate observed patterns of host preferences in herbivore populations. Thus only genetically based variation can serve as raw material for natural selection to bring about host shifts, expansion of host range, or divergence in host use by different populations (see Futuyma, Chapter 8, this volume). By contrast, both genetic and nongenetic sources of variation could conceivably be exploited by humans to alter an insect's host-location behavior in a desirable manner. For example, breeding of host specificity for a weedy plant may be expected to improve the efficacy of a herbivorous insect species as a biological control agent for that plant. If adult conditioning occurs in that insect, however, effectiveness might also be significantly enhanced by exposing adults to the target species before release—a program that could be much less expensive and time consuming than a breeding program. Which of the two approaches toward control should be adopted will thus depend on the relative importance of genetic variation and adult conditioning as sources of variation in host-finding behavior in the insect under consideration. In general, the methods adopted for manipulating behavior will be governed by knowledge of the mechanisms responsible for variation in that behavior.

In this chapter, attention will be focused on those environmental sources of variability in host-seeking behavior that can account for variation in host use within and between phytophagous populations. In doing so, we emphasize that host selection is a catenary process, a chain of behavioral responses each of which is contingent on the perception of stimuli arising as the result of a previous response (Kennedy, 1965a); the earlier steps in the sequence comprise host-location behavior. Hence, any two searching insects may differ not only in the type of response (e.g., taxis versus kinesis and klinokinesis versus orthokinesis), the level of response (e.g., velocity of linear movement and rate of turning), or the probability of *any* response to host-plant stimuli. They may also differ in the way in which responses made early in the host-finding sequence influence responses made later in the sequence or in succeeding sequences, phenomena best exemplified by the effects of experience and antagonistic reflexes on host-seeking behavior. The discussion repeatedly stresses the inadequacy of studies of behavioral determinants of host use that center on a limited series of stimulus–response pairs to the exclusion of the rest of the host-selection sequence. In this regard, our notions of the behavioral mechanisms underlying specialization by phytophagous insects have been derived almost entirely from investigations of discrimination subsequent to host discovery. This shortcoming has not only underplayed the significance of host-location behavior in determining patterns of host use, but has also led to some inappropriate views on the evolution of host-selection behavior.

In addition to reviewing evidence for environmentally induced variation in host-location behavior, this chapter addresses ways in which genetic variation may be employed to analyze the neurological underpinnings of host-selection behavior. Finally, it provides a paradigm based on the authors' own research for determining how environmental and genetic factors contribute to the behavioral repertoire associated with host location by a phytophagous insect in which variation in host use is maintained by variation in host-seeking behavior.

## II. ENVIRONMENTAL SOURCES OF INDIVIDUAL VARIATION IN HOST LOCATION

A diversity of environmental factors can influence host-location behavior, but only some of these are known to alter host selectivity by searching individuals. Special consideration has been given here to those environmental factors that have been shown to or could conceivably play a role in establishing and maintaining host preferences between or within phytophagous insect populations. Discussion of other effects such as circadian or other biological rhythms has been omitted, although these factors assuredly produce variation in host-finding behavior. In any case, recognition of all environmental effects is of operational significance in the genetic analysis of host-seeking behavior. Heritability and response to selection are among many quantitative genetic measures that are most accurately estimated when environmental effects are carefully controlled (Falconer, 1980).

Environmental sources of variation can be characterized according to their physiological mode of action on the behavior of the individual. Experience, for example, influences behavior through immediate and reversible changes in the nervous system. Other effects, like photoperiod, appear to exert humoral effects during morphological development that can permanently alter the ''hard-wiring'' of response mechanisms. Still others have impacts at several physiological levels. Ambient temperature, for example, can change behavior momentarily by modifying current metabolic rate, or it can act as a cue regulating developmental pathways that permit little subsequent modification of behavior.

We devote little attention here to the details of the physiological processes through which these factors act to alter behavior. Most of the physiological mechanisms are not well understood and those that are known have been reviewed elsewhere (Rankin, 1978; Truman, 1978; Nijhout and Wheeler, 1982). Nevertheless, it is recognized that knowledge of both the time scale over which these processes act to modify discriminatory behavior and the resulting reversibility of changes in host selectivity is crucial to predicting ensuing patterns of variation in host use. Moreover, the success of a biological control program like the one just proposed may rest on the durability of environmental effects on host-selection behavior. Our objective in this section is to review evidence of environmental sources of individual variation in host-finding behavior and, where possible, to illustrate the ways in which this variability can alter host use in herbivore populations.

## A. Experience

### *1. Previous Examples among Phytophagous Insects*

Although the searching behavior of many insects is modified by experience (von Frisch, 1967; Alloway, 1973; Menzel *et al.*, 1974; Dethier, 1976), examples among phytophagous insects are few. Despite this paucity of information, variation in host preference among phytophagous populations may reflect the varying histories of individuals within those populations and may be maintained by effects of experience on host-location behavior.

Most investigations into the effect of experience of individuals on patterns of host use by populations have focused on the Hopkins Host Selection Principle and so-called induction of preference. Whether intended or not, these studies by virtue of their design examined only aspects of discrimination subsequent to host discovery. Nevertheless, because both phenomena presumably could impinge on host preference through influences on host-location behavior, the evidence for each is briefly reviewed.

The well-known Hopkins Host Selection Principle asserts that adult individuals tend to feed or reproduce on the food-plant type they consumed during the juvenile stages (Hopkins, 1917). Since this hypothesis was proposed, it has received some limited experimental support (Hovanitz and Chang, 1963; Hovanitz, 1969; Yamamoto *et al.*, 1969; Phillips and Barnes, 1975; Hsiao,

1978). Although such effects have almost always been attributed to conditioning, the principal objection to that conclusion is that these studies have uniformly neglected to control for possible artificial selection for those genotypes favoring the host on which they were reared (Jermy *et al.*, 1968). Two conditions would favor artificial selection in these experiments: (1) genetic variation in host selectivity and host-specific survival, and (2) a genetic correlation between ability to survive on a host type and tendency to select that type. In the event both sufficient genetic variation and the specified genetic correlation exist, then those individuals that survive on a host are more likely to prefer the host type on which they were reared (e.g., S. Via, unpublished). The data of Phillips and Barnes (1975) in particular suggest such a component of artificial selection. The shift in selectivity of a population of the codling moth *Laspeyresia pomonella,* which originally favored walnut, did not appear until the second generation of adults that were reared on apple. Artificial selection (rather than some unusual cumulative maternal effect) was probably responsible. In any case, the large body of evidence accumulated against the influence of larval diet on adult preference (Wiklund, 1974; Phillips, 1977; Claridge and Wilson, 1978; Copp and Davenport, 1978; Stanton, 1979; Tabashnik *et al.*, 1981) has cast great doubts regarding the general significance of the Hopkins Host Selection Principle in determining observed variation in host preference in phytophagous insect populations.

Conditioning of preference within one developmental stage, by contrast, has been observed almost everywhere it has been examined (Jermy *et al.*, 1968; Wiklund, 1973; Phillips, 1977; Cassidy, 1978; Barbosa *et al.*, 1979). In most cases, the larval stage is affected, but adults are occasionally susceptible to conditioning as well (Phillips, 1977; Cassidy, 1978). Although rarely tested, induction of preference is at least sometimes reversible (e.g., Cassidy, 1978; cf. Jermy *et al.*, 1968). The modification is apparently neurally based (Hanson and Dethier, 1973), which suggests that it is indeed a type of learning. The term by which this phenomenon has been denoted, induction of preference, is somewhat misleading when applied to every occurrence (although it was appropriately used by Jermy *et al.*, 1968). Frequently, there is a preconditioning preference, and rearing on a single food type simply strengthens or weakens that preference. A preference is induced only when a preference results wbere none previously existed (Wiklund, 1973), or when the rank order of preferred plants changes as a result of experience with a host plant. A change in the strength of a preference may more appropriately be referred to as *modification of specificity* and a change in the order in which plant types are selected simply labeled as a *modification of rank order*. These terms have an immediate affinity with the concepts introduced by Singer (1982b) and discussed in Section II,B. Whether these distinctions characterize unique physiological mechanisms or have different ecological ramifications remains to be investigated, but the assumption that they do not seems premature.

*2. Classification of Learning Types*

Learning, which can be defined as a change in behavior with experience (Alloway, 1973), has traditionally been partitioned into several categories: habituation; associative learning, which consists of Type I (or classical conditioning) and Type II (including trial-and-error learning, instrumental learning, and operant conditioning); latent learning; and insight learning (Thorpe, 1963). This classification is derived from descriptions of behavior observed under controlled schedules of the timing and the rate of reinforcement of stimuli. They do not imply an analogous division of underlying neural mechanisms (Dethier, 1976). It is difficult to distinguish among the types under most field situations and, in fact, most laboratory studies on learning in insects do not adequately distinguish among them either (Dethier, 1976). Learning, for instance, has been presumed for foraging insects that repeatedly return to rewarding nectar sources or host plants (Janzen, 1971; Ehrlich and Gilbert, 1973; Gilbert, 1975; Frankie *et al.*, 1976). If so (and there is no direct evidence that it is learning), it is impossible to say whether a simple form of learning, such as habituation or a more complex type such as associative learning, is responsible for such "traplining." A female *Heliconius* butterfly traplining from one *Passiflora* plant to another (Gilbert, 1975) may do so because it has become habituated to all environmental stimuli associated with searching in unprofitable areas. Alternatively, the insect may trapline because it has learned to associate certain local landmark stimuli with the reward contingent on successfully ovipositing on a *Passiflora* plant. Associative learning may seem the obvious mechanism, but in order to distinguish rigorously between the two learning types, it is necessary to correctly specify and manipulate all of the relevant stimuli, a quite unlikely prospect for a field biologist!

Previous attempts to fit insects into traditional learning categories have produced variable and obscure results (Alloway, 1973; Dethier, 1976). Therefore it appears doubtful that this paradigm can contribute many insights into the effects of experience on the host-selection behavior of phytophagous insects. Just as previous investigation has failed to uncover any apparent phylogenetic patterns in the occurrence and development of these learning types (Dethier, 1976), it is also unlikely that this classification will shed any light on the evolution of host preferences among phytophagous insect populations.

We suggest an alternate classification of learning that is perhaps more pertinent to the ecology of phytophagous insects and that, at the same time, probably reflects a relatively discrete partitioning of neural mechanisms as well. We propose to specify learning according to the position in the stimulus–response chain that is modified by experience: *conditioning of response* when the probability of an insect exhibiting a certain response increases with repeated presentation of stimuli; and *conditioning of perception* when the probability of perceiving a stimulus or stimulus complex increases as a result of repeated presentation of stimuli. We contend that such a classification will eventually be more useful in predicting host use in a herbivore population whose members learn to search for and select host plants. As inferred later, the efficacy of conditioning of response is likely to depend on the densities, distributions, and quality of available host types. By contrast, the extent to which phytophagous insects employ conditioning of perception will hinge on the density, spatial pattern, and diversity of the plant community in which host types are found. Thus a distinction is made between attributes of the host populations and components of the resident non-host community each of which are already known to influence herbivore search patterns in different ways (see Stanton, Chapter 4, this volume).

*3. Conditioning of Response*

Most examples of insect learning under laboratory conditions probably involve the conditioning of a certain response with repeated presentation of stimuli. The usual laboratory assay uses an environment that is free of any competing stimuli, so that all stimuli are perceived when encountered. In this setting, the probability of detection is effectively unity and cannot be improved by conditioning. As such, the results of these studies probably reflect conditioning of response. These laboratory studies on nectar-seeking insects have produced a number of generalizations of possible relevance to the study of the effects of experience on host location in phytophagous insects.

1. When two stimuli or stimulus complexes are associated with different rates of reward, the insect chooses the one with the higher rate, all other things being equal (Menzel and Erber, 1978). Phytophagous insects might likewise learn to search for the host-plant type associated with the higher rates of feeding, oviposition, or host discovery.
2. The rate of learning (expressed in terms of the numbers of trials to peak responsiveness), is not influenced by the quantity or quality of the reward per trial (Menzel, 1968; Menzel and Erber, 1972). For phytophagous insects, this suggests that the rate of learning would not be influenced by the quality of the host plants on which they are feeding or reproducing.
3. The rate and peak responsiveness of learning is contingent on the physical nature of the reinforcing stimulus. Among honeybees, for example, olfactory cues are learned more quickly than visual cues (Menzel, 1967; Robacker and Ambrose, 1979). Heinrich *et al.* (1977) noted that bumblebees more quickly associated sucrose rewards with blue artificial flowers than with white ones in enclosure experiments conducted against a uniform green background. If applicable to searching phytophagous insects, variation in the kinds of cues emitted by host-plant types may lead to differential modification of responses to those

host types. A host preference in a herbivore population may be generated that is not directly related to the relative rates of feeding or reproduction associated with exploiting the available host-plant types.

4. Insects are more likely to be conditioned to stimuli associated with continuous reward schedules than those associated with intermittent reward schedules. Given a choice of two types of artificial flowers randomly arrayed in enclosures, bumblebees learn to alight more frequently on the flower type associated with the lower variance in the reward per flower even when the mean reward for flower types is equivalent (Waddington *et al.*, 1981). On this basis, we might expect phytophagous insects to learn to search for the host type that is most uniform in quality among host individuals.

5. Learning is usually reversible (Menzel and Erber, 1978). Thus individual insects responding more frequently to signals from one plant type might occasionally switch to respond more frequently to the other(s). The change in the differential response to alternate host plants might be the result of a change in reward rates (i.e., rates of feeding or oviposition) on the alternate types. However, even if the relative reward rates do not change but are similar, sampling error by the searching individuals may occasionally induce a switch.

Moreover, the tendency of individual insects to switch depends on the physical nature of the cues. Heinrich *et al.* (1977) trained bumblebees to search for blue or white artificial flowers arrayed in enclosures by supplying one or the other flower type (but never both simultaneously) with sucrose solution. After the initial training period, sucrose rewards at the conditioned flower type were stopped and the other flower type was supplied with sucrose solution. Individuals trained to blue flowers switched less readily to white than bumblebees trained to white flowers switched to blue. Apparently, bees majoring on blue visit the alternative flower type less frequently than do bees majoring on white. Hence, the shift in relative reward rate is assessed later in bees foraging on blue flowers. This result is consistent with the finding that color retention in honeybees is longer for those colors that are more quickly learned (Menzel *et al.*, 1974). Likewise, some phytophagous insects that can shift to alternate host types might nevertheless exhibit a lag period between the change in reward rates and the change in responses to alternate host types. The duration of "constancy" to the previously favored host may depend on the quality of cues emitted by that host type.

6. Insects appear to have several types of memories that vary in their temporal stability. In the first few seconds following the perception of a color signal by a honeybee, information is stored in a "sensory" memory. If the signal is reinforced (e.g., sucrose reward), it will subsequently be incorporated in a "short-term" memory lasting about 5–7 min. Eventually, barring interruption of consolidation by environmentally induced trauma or learning of alternate signals, the signal is retained in a "long-term" memory that is quite impervious to decay (Menzel *et al.*, 1974).

Extrapolation of any of these last laboratory studies on bees to field observations on phytophagous insects must be done with caution. Nevertheless, some recent evidence on learning by ovipositing butterflies is concordant with the memory-consolidation process illuminated by Menzel and colleagues (Menzel, 1967, 1968; Menzel and Erber, 1974, 1978; Menzel *et al.*, 1974). Female butterflies (*Colias* spp.), searching for host plants on which to lay their eggs, alight on leaves of many plants, some hosts and some nonhosts. In a 5- to 20-min observation period, M. L. Stanton (unpublished) found that individuals of several species of *Colias* butterflies make proportionately fewer alightings on nonhost plants as an oviposition bout progresses. Stanton also noted that intervening nectaring visits to flowers reversed the learning process and the number of landing "errors" increased at the beginning of the next oviposition bout. Although the typical length of an oviposition period between nectaring sequences was not specified, it is conceivable that they were shorter than required for commitment of the learned response to long-term memory and were subsequently extinguished. In those phytophagous insects like *Colias* butterflies that must divide their time among several kinds of searching activities, there may be fundamental physiological limits to learning. This should not, of course, discourage the pursuit of studies on learning and host location in phytophagous insects. Indeed, the remarkable discovery by M. L. Stanton (unpublished) is that learning plays *any* role in the host-location behavior of phytophagous insects. As described later, her demonstration of the effect of experience on host-finding behavior has corroborative support in our studies on pipevine swallowtail butterflies. In the case of the pipevine swallowtail butterfly *Battus philenor*, variation in individual experience contributes to variation in host use directly through influences on host-location behavior.

*4. Conditioning of Perception*

Conditioning of perception refers to any effect of experience that improves the probability of detecting a stimulus or a stimulus field. The most popular conception of conditioning of perception is search image formation. First elaborated by Tinbergen (1960) to explain patterns of foraging behavior by insectivorous birds, evidence for the existence of search images has been repeatedly reported for vertebrate foragers (for review of these and others see Krebs, 1973; Pietrewicz and Kamil, 1979). Unfortunately, the term has acquired a number of meanings by these and other investigators, including area-restricted search (Croze, 1970) and even simple preference (Atema *et al.*, 1980). As originally put forward, a search image involves learning to detect the object of search, that is, the animal "learns to see" the item of search. Such a perceptual refinement contrasts with conditioning of response in which the animal "learns to prefer" some food item. A search image has several implied properties besides enhanced perception: (1) a correlated decrease in the probability of detecting alternate target types, (2) enhancement of discovery rates for the target (but see Croze, 1970; Murton, 1971), and (3) reversibility. Search images have also been associated with a generalization of complex sets of stimuli, that is, a perceptual filtering of the array of the cues emitted by the target (Tinbergen, 1960; Hinde, 1970). Nevertheless, behavioral observations of searching insects that appear to respond to only a subset of the cues emanating from a flower or a host plant do not constitute sufficient evidence for conditioning of perception (or even conditioning!), even though they have sometimes been described as search images (e.g., Swihart and Swihart, 1970; Wiklund, 1974).

In fact, it is somewhat difficult to distinguish between search image formation and conditioning of response on the basis of behavioral observations alone. Both mechanisms may increase discovery rates. Both may result in search biased toward one food type when several are available in the habitat. The extent to which either mechanism is employed toward search for a particular target is likely to depend on the reward rates associated with alternate target types in very similar ways.

Despite these similarities, there is one significant difference that has proved operationally useful in distinguishing between the two kinds of conditioning. Search images are most effective when the target is cryptic; the efficacy of conditioning of response, by comparison, should not vary with the degree of crypticity of the target. If increases in discovery rates with experience are observed only under cryptic conditions, adoption of a search image is the more likely mechanism (Dawkins, 1971a,b; Pietrewicz and Kamil, 1979). For phytophagous insects, the crypticity of host plants depends on the characteristics of the nonhost community in which host types are embedded. These characteristics include the density, diversity, and distribution of the vegetation surrounding host plants, all of which are believed to influence the success of insect search and the resulting loads on various host types (Stanton, Chapter 4, this volume). Nonhost vegetation frequently masks host plants from discovery by herbivorous insects (e.g., Rausher, 1981a). Host plants are thus often cryptic to searching insects, and herbivore searching efficiency may potentially be ameliorated by search image formation.

In experiments using cryptic conditions to discern search image formation, a great deal of attention has been given to controlling for effects of area-restricted search, familiarity with the target, innate preferences for certain target types, and alterations in handling time or search path due to experience (Krebs, 1973). Each of these factors can potentially cause increases in discovery rates with recent encounters of the target type that are not related to formation of a search image. These considerations have culminated in several experiments that show rather unambiguously that vertebrate foragers are capable of forming search images (Dawkins, 1971a,b; Pietrewicz and Kamil, 1979). Yet, although Dethier (1978) suggested that the notion of search images and the techniques devised by vertebrate biologists offered numerous research possibilities for students of phytophagous insects, there is no evidence to date for search image formation in any invertebrate. By the criteria just outlined, the claim by Rausher (1978) to have demonstrated search image formation in *Battus philenor* was premature.

### B. Time-Dependent Responsiveness

In 1922, Knoll reported that the specificity for oviposition sites by female sphingid moths (*Macroglossum stellatarum*) declined if females were prevented from laying (cited in Hinton, 1981). At first females tended to approach and land on objects that exuded host-plant odor. If they were prevented from laying eggs for a time, however, eggs were eventually deposited on anything with the appropriate color. Once egg laying commenced, the relaxation in stimulus specificity was reversed.

Observations like Knoll's have traditionally been associated with the notion of motivational states. Unfortunately, the term *motivation* has been used and misused in a variety of ways. Hindered by obfuscating anthropomorphism, the concept has little to recommend it for any animal (Dethier, 1976; but see Alcock, 1979). Nevertheless, reports by Knoll and others suggest that temporal variation in responsiveness to host cues is common among phytophagous insects. Such time-dependent responsiveness, as we shall refer to the phenomenon, has the following properties: (1) the probability of an observable response to host stimuli (i.e., the responsiveness to host stimuli) increases with time elapsed since performance of a consummatory act such as feeding or oviposition; (2) the change in responsiveness with time is independent of prior exposure to host stimuli or performance of other activities besides host seeking (and presumably reflects some internal physiological variable that changes monotonically through time); and (3) the change in responsiveness is reversible upon performance of a consummatory act.

Time-dependent responsiveness to host cues can affect multiple host use in phytophagous insect populations if unique time periods are required before particular host types can be utilized by individual insects. Singer (1982) and Rausher (1983) have extended this notion to develop quantitative measures of host preference by ovipositing checkerspot butterflies (*Euphydryas editha*). Their model assumes that to each potential host species (or other category of plants that females can discriminate), there corresponds an acceptance threshold. Females will oviposit on an encountered plant species only when the time elapsed since last oviposition exceeds the threshold for that species. Singer's assay in particular provides two quantitative measures of host specificity: (1) the rank order of selectivity, that is, the order in which host types become acceptable over the time since eggs were last laid; and (2) the specificity of selectivity, that is, the rate at which host types become acceptable over the time since eggs were last laid. For the simple case of two alternate host-plant species, Singer defines several phases through wbich a butterfly progresses in the time since last oviposition (Fig. 1). There is an initial refractory period after egg laying during which the butterfly will not oviposit on any plant. During the subsequent "discrimination phase," the insect accepts one plant species for oviposition, but not the other. If the animal has not yet oviposited, this phase is eventually followed by one in which both host species become acceptable.

Singer defines specificity as the length of the discrimination phase. In Fig. 1, individuals in population A have longer discrimination phases than members of population B. Population A is therefore expected to be more host specific than population B because A individuals are more likely to discover and oviposit on Host Type 1 before the time elapsed since prior oviposition exceeds the acceptance threshold for Host Type 2 than are B individuals.

Singer's definition of specificity does not take into account the refractory period of an ovipositing butterfly. Yet, in some circumstances, differences in the lengths of the refractory periods might lead to differences in host use. As an example, consider the acceptance thresholds of individuals in populations A and C in Fig. 1. The discrimination phases of individuals in populations A and C are

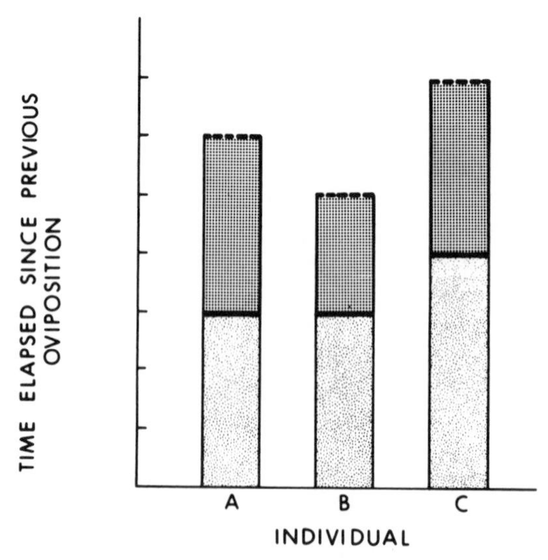


**Fig. 1.** Positions of acceptance thresholds and corresponding lengths of refractory periods and discrimination phases for two host types for three ovipositing individuals representative of each of three hypothetical herbivore populations whose members exhibit time-dependent selectivity. Solid lines, acceptance threshold for Host Type 1; dotted lines, acceptance threshold for Host Type 2. Lightly dotted bars, refractory period; boldly dotted bars, discrimination phase.

equivalent, but members of population A have a shorter refractory period than members of population C. By Singer's criterion, both populations should be equally host specific. If the longer refractory period of individuals in population C significantly increased the probability of discovering Host Type 1 during the discrimination phase, however, population C might be more host specific than population A. Such would be the case if Host Type 1 was regularly spaced and rare enough that individuals in population C tended to discover the first plant of Type 1 (since the previous oviposition) during the discrimination phase, whereas members of population A tended to discover the first plant of Type 1 after the discrimination phase had passed. The resulting specificity of either population would also depend on the abundance and spatial pattern of Host Type 2 (as well as an array of other factors including discrimination among conspecific hosts after discovery), but the point remains: host specificity in phytophagous populations whose members exhibit time-dependent selectivity can depend on the lengths of both the discrimination phase and the refractory period.

In any case, specificity and rank order of selectivity provide two discrete, measurable behavioral characters by which specialization may be quantitatively assessed. We add parenthetically that Singer assayed postalighting discrimination more rigorously than most previous attempts by carefully manipulating the exact sequence of encounters with the alternate host species. In other words, if insects in small cages cannot exhibit normal host location, neither are they likely to display natural postalighting behavior. Thus confined, the insect's inability to search normally will lead to irregular rates of encounter with the host plants. This, in turn, will result in irregular levels of discrimination and these will precipitate irregular patterns of postalighting selectivity. In lieu of natural sequences of encounters, Singer's technique of random exposure to the host-plant types controls for these historical effects.

Although Singer's observations on checkerspot oviposition behavior are restricted to examination of postalighting behavior, his ideas should be applicable to discrimination before host discovery as well. Desert locusts (*Schistocerca gregaria*), for example, that are progressively starved become increasingly likely to fly upwind toward grass odors under wind tunnel conditions (Haskell *et al.*, 1962). This upwind response to food odors decreases after contact with food in proportion to the length of time exposed to a food source (Moorhouse, 1971). In all cases, renewed deprivation is again accompanied by a gradual increase in upwind responses over time. Similar effects of resource deprivation on searching behavior in other phytophagous insects (Dingle, 1968; Saxena, 1969, 1978; Bernays and Chapman, 1974; Barton Browne, 1975) suggest that host-location behavior commonly changes with time elapsed since feeding or reproduction. Whether time-dependent responsiveness during host seeking is even partly responsible for observed host specificity in herbivore populations is, however, presently unknown.

## C. Event-Dependent Responsiveness

An individual's responsiveness to host-finding stimuli may be modified by prior behavioral episodes unrelated to host selection, a phenomenon that might be described as a kind of event-dependent responsiveness. Furthermore, host preference in a herbivore population may be tied to its members' performance of activities other than host seeking if some host types are more likely than others to elicit or inhibit such behaviors.

Migratory behavior, for example, is particularly interrelated with host-location behavior. Migration has been defined as persistent, straightened-out movement that is accompanied by internal inhibition of vegetative responses that will eventually arrest the movement (Kennedy, 1961a). Indeed, Kennedy and others have urged that a principal criterion for migration be low responsiveness to vegetative stimuli, which include those associated with feeding, mating, and reproduction. We might therefore expect that variation in the tendency to migrate will be accompanied by variation in the tendency to search for hosts actively. Likewise those external cues or internal physiological factors that stimulate migratory behavior are likely to be those that inhibit host-location behavior. If migratory behavior is a function of the age of the individual (Dingle, 1972), host location is likely to be a function of age as well. If migratory flight polymorphisms exist in many insect populations (Dingle, 1972; Rankin, 1978), such populations are probably polymorphic for host-location activity. In short, migratory insects might be exceptionally profitable subjects for the study of variation in host-location behavior. Unfortunately, Kennedy's criterion for migration has seldom been applied in practice, especially with respect to vegetative stimuli pertinent to host selection (even though the majority of insect migrants are phytophagous). When and if such criteria are rigorously applied, we are likely to see that there is a continuum between migration and so-called trivial or appetitive flight (Dingle, 1968; Kennedy, 1975; Rankin and Rankin, 1980), revealing more complex patterns of variation in behavior than is implied by the present dichotomy of classification.

Although migratory behavior often inhibits behavior associated with host finding, it frequently primes the individual for future responsiveness to vegetative stimuli. Long periods of flight hasten the onset of oviposition in migratory insects such as the milkweed bug (Slansky, 1980) and, presumably, the initiation of the host-location behavioral sequence. Migratory flight and host-selection behavior represent closely linked antagonistic reflex systems in the black bean aphid, *Aphis fabae* (Kennedy and Booth, 1963; Kennedy, 1965b, 1966; Kennedy and Ludlow, 1974). Stimulation of one reflex system inhibits the other; thus, for example, host settling is deterred by elicitation of flight toward a light source. In addition, inhibition of a reflex by its antagonist is followed by an aftereffect on the inhibited system. This residual rebound is characterized by either compounded inhibition (antagonistic depression) or stimulation (antagonistic induction) of the reflex. The balance between induction or depression depends both on the internal states of each system and on the sensory inputs received by the aphid (Kennedy, 1966). Continuous flight uninterrupted by presentation of settling stimuli, for instance, increases the likelihood of a settling response when a target is next presented. Repeated and brief periods of settling, by contrast, will result in extended flight and reduced responsiveness to subsequent settling stimuli.

Antagonism between migratory flight and host-selection behavior in the black bean aphid is not restricted to discrimination subsequent to host contact. Kennedy and Ludlow (1974) have shown that migratory flight (movement toward a light source) and targeted flight (movement toward a source of host-settling stimuli) interact in a similar fashion. Performance of migratory flight primes the aphid for approach to a yellow hostlike target. Repeated presentations of the target without allowing settling establishes a pattern of continuous migratory flight. As in the flight–settling systems, the degree of induction or depression of either system is determined by prior events in the behavioral sequence.

The key point is that the responsiveness of individual insects to cues associated with finding the host plant is partly a function of the insect's recent history of exposure to those cues and of the performance of activities other than host selection. If cues from various host types differ in the tendency to induce an activity or to prime the individual for responsiveness to cues from host plants subsequently encountered, specialization on particular host types might be affected. Host preference in a population whose members exhibit antagonistic induction of migratory flight by settling, for example, could be markedly altered if different host types occur in more or less discrete patches. Under these conditions, a given insect is likely to be successively exposed to cues from the same host type. Differences between host types in the tendency to prime host settling will lead to differences in patch residence time even if patch sizes and host density within patches are similar among host types. Furthermore, significant asymmetries in herbivore loads on the available host types could result even when insects exhibit little postalighting selectivity among host types in simple choice tests.

In addition, Kennedy's models of antagonistic reflex systems provide a behavioral mechanism for negative relationships between host density and the percentage of plants attacked within patches (e.g., R. B. Root and P. Kareiva, unpublished; see also Stanton, Chapter 4, this volume). High-density patches (which signify high rates of stimulus presentation to an insect assumed moving at the same speed and directionality in all patches) may induce extended flight, reduced responsiveness to host stimuli, and shorter patch residence times, all of which might more than compensate for increased encounter rates within those patches. Host-preference patterns in the population may also be influenced by this process if patches of different host types tend to differ in average density.

These scenarios are speculative to be sure, but they serve to underscore the heuristic value of considering the consequences of multiple host encounters on subsequent host-location behavior.

## D. Host-Plant Contact

The priming or depression of future responsiveness to host cues noted in the previous section is only one of numerous ways in which exposure to a host plant can modify host location by phytophagous insects. Contact with a host plant can have long-term developmental *primer* effects on the host-finding activity of its herbivores as well as immediate, short-term, and apparently neurological *releaser* effects on subsequent host-location behavior. Both primer and releaser effects can create individual variation in behavior and, if their influence on host location varies among host types, both can mold host preferences in phytophagous insect populations.

Regarding primer effects, host-plant characteristics often regulate the onset of oogenesis and associated oviposition behavior. Food quality determines the length of the adult prereproductive period in the female diamondback moth *Plutella maculipennis* (Hillyer and Thorsteinson, 1969). Apparently, contact or close proximity to the host plant stimulates the ovaries, influencing the onset of oviposition independent of mating. Feeding induces oocyte formation and flight muscle histolysis in three African species of cotton stainer bugs (*Dysdercus* spp.) (Dingle and Arora, 1973), events that are associated with the cessation of migratory flight and the initiation of oviposition. Seed feeding by female milkweed bugs (*Oncopeltus fasciatus*) is necessary to complete oogenesis; poor host quality results in delayed reproduction and enhanced migratory activity (Rankin, 1974). The presence of bean seeds stimulates oogenesis in the bean weevil *Acanthoscelides obtectus* (Pouzat, 1978). Food intake also initiates oogenesis and oviposition in the Colorado potato beetle *Leptinotarsa decemlineata* (de Wilde *et al.*, 1969). Whether feeding or other host contact has more profound influences on aspects of host location in these species besides the onset of host-finding behavior is presently unknown.

Releaser effects of host-plant contact include temporary and immediate changes in the probability of a host-seeking response. The adult female cabbage butterfly, *Pieris rapae*, for example, lands more frequently on materials possessing the appropriate visual stimuli after landing on host plants. Heightened responsiveness to leaf disks elicited by repeated contact with the host plant is, in this case, not associated with induction of ovarian development (Traynier, 1979).

Host-plant contact may also modify host-seeking responses in a graded fashion. Discovery of a food item often immediately increases the rate of turning of a foraging animal (Banks, 1957; Dethier, 1957; Dixon, 1959; Kaddou, 1960;

Chandler, 1969; Croze, 1970). Such a change in searching behavior tends to increase discovery rates when the target has a patchy distribution. Similar alterations in movement patterns have been observed among phytophagous insects searching for host plants. The aphid *Brevicoryne brassicae* increases the number of short flights in the vicinity of a cabbage host plant recently encountered (Kennedy *et al.*, 1959b). Similarly, ovipositing *Cidaria abulata* moths turn more frequently in response to the odor of their host plants (Douwes, 1968). Cabbage butterfly larvae (*Pieris rapae*) also exhibit reduced directionality after contact with a host plant (Jones, 1977a). This behavior may not be limited to the larval stage of cabbage butterflies. Adult female cabbage butterflies also increase turning rates after contact with hosts under field conditions (Jones, 1977b). Root and Kareiva (in press) recently obtained conflicting results for ovipositing *P. rapae*. Females observed by Root and Kareiva were highly directional after oviposition. The disparity may be attributable to the scale of distance from the host plant over which directionality was assessed in each case. Butterflies do flutter briefly around a host on which they have laid an egg; soon after, however, flight is highly directional (P. Kareiva, personal communication). Root and Kareiva further assert that this overall directional response ensures that eggs are spread over large areas and thus reduces the variance in survivorship due to localized sources of juvenile mortality.

In terms of selective pressures, this advantage of risk spreading acts in direct opposition to the tendency to increase discovery rates in patches by increasing turning rates (Tinbergen *et al.*, 1967). Thus far, however, host-induced increases in turning rates have never been explicitly correlated with increased host-discovery rates by phytophagous insects. Nor has anyone shown for phytophagous insects a phenomenon similar to that shown in houseflies in which the change in turning rate is in proportion to the quality of the food item encountered (Dethier, 1957). Moreover, if the magnitude of the change in turning rate is unique for each host type, patterns of host use could be significantly altered, particularly when host types are distributed in discrete patches. Under those conditions, herbivores may remain longer in patches of host types that elicit higher turning rates, all else being equal. Ovipositing cabbage butterflies (Jones's populations!), which increase their turning rate subsequent to discovery and which discriminate among host plants of different ages and varieties (Ives, 1978; Jones and Ives, 1979; Latheef and Irwin, 1979), might provide an appropriate system in which to study these effects.

Search speed is another orientational parameter on which host-plant contact has graded releaser effects. Cabbage butterfly larvae, for example, move more slowly in a given direction after discovery of a host plant (Jones, 1977a). Although it has been suggested that a reduction in linear velocity may enhance the probability of detecting an encountered food item (Jones, 1977a; Gendron and Staddon, in press), this remains to be shown empirically. Whether host-specific effects on search speed can alter host preferences in herbivore populations, too, has yet to be demonstrated rigorously. Nevertheless, Stanton (1982) has provided some indication that search speed in *Colias* butterflies is related to the host type most recently encountered. Median flight velocity among ovipositing *Colias philodice eriphyle* individuals depends on the host species last visited. Moreover, displacement rate after contact with five leguminous host species roughly paralleled the probability of oviposition after landing on these host species. Given that all host species are patchily distributed, slower velocities for host types associated with higher probabilities of oviposition after landing ought to increase attack rates most for those host types that constitute the most favorable oviposition sites. In this way, herbivore loads on these species might exceed those predicted by an examination of postalighting discrimination alone.

Stanton's search velocity estimates represent the square of linear displacements that are measured over variable periods of time. Thus different measured velocities might reflect a disparity in instantaneous linear velocities or in turning rates. Despite this ambiguity, graded releaser effects can potentially alter patterns of oviposition preference in *Colias philodice eriphyle*. Numerous other examples of graded effects of host quality and type on dispersal behavior (Dixon, 1969; Capinera and Barbosa, 1976; Kidd, 1977) recommend further examination of the consequences of host contact for patterns of host use.

Finally, responses to aspects of host-plant distribution such as degree of patchiness, density, and microhabitat can be an additional source of individual variation in observed host-location behavior, but are reviewed elsewhere (Stanton, Chapter 4, this volume). From the standpoint of both precisely specifying behavioral mechanisms and accurately predicting impact on host preference, we remark that attention should be given to distinguishing between a dependence of host-finding behavior on plant distributions that is a statistical result of summing effects of individual host plants (e.g., Jones, 1977b), and an actual differential response to alternate plant distributions (e.g., Douwes, 1968).

### E. Age-Related Parameters

Age is a somewhat loosely defined parameter, referring sometimes to time since hatch, sometimes to time since molting to the adult stage, and sometimes to time since mating. In experiments, age is frequently confounded with other temporal patterns such as food-deprivation periods (see Barton Browne, 1975), physiological cycles, or history of exposure to food plant. Furthermore, age-related variation may be rooted in developmental changes that are internally programmed or externally regulated by environmental cues or in neural refinements associated with experience. Yet the properties of such changes are radically different. The permanence and reversibility of behavioral modifications, for example, is probably contingent upon the mechanism underlying the change.

Still, for all the attendant ambiguity, host-location behavior varies with the age of the individual in a variety of phytophagous insects.

Oviposition rates in many insects, for example, rise above zero at some point after mating, increase daily to a maximum, and decrease subsequently to zero or until death (Labine, 1968; Dunlap-Pianka *et al.*, 1977; Gossard and Jones, 1977; Ichinosé and Honda, 1978). This variation in oviposition rates with age may be manifested as variation in either life history characteristics or host-selection behavior. For example, lower oviposition rates in older individuals may be reflected in lower average egg cluster sizes per oviposition. Lower oviposition rates may be accompanied instead by a rejection of a larger proportion of discovered plants as suitable sites of oviposition. Alternatively, age specificity in oviposition rate may be expressed as variation in features of host location such as time spent searching, sensitivity to host-plant or habitat cues (including either thresholds of perception of cues or responsiveness to those cues perceived), and search speed. Despite the prevalence of age-related variation in oviposition rates and the simplicity of these alternate hypotheses, experiments differentiating among them are few. Nevertheless, there is some indication that responsiveness to chemotactile host-plant stimuli wanes as some insects age. Ichinosé and Honda (1978), for instance, remark that older female *Papilio protenor demetrius* butterflies (which typically have lower oviposition rates than freshly mated individuals) exhibit fewer consistently positive oviposition responses to host-plant extracts than young individuals.

Host location clearly changes with age in a variety of other phytophagous insects. Among some macropterous leafhoppers, flight ability (and presumably host-finding ability) varies with age (Waloff, 1973). Among sycamore aphids (*Drepanosiphum platanoides*), the tendency for winged morphs to fly and find new hosts in the forest canopy declines with age (Dixon, 1969). Similarly, average flight duration of milkweed bugs (*O. fasciatus*) decreases with age; older individuals also exhibit an increased tendency to fly in laboratory assays (Dingle, 1965). The pattern of change with age in both flight parameters is sex specific: males show a second peak in flight duration as they age that is absent among aging females. Also, some older males have markedly higher flight thresholds than any female of similar age. Dingle characterizes these short flights as trivial flights associated with changing of food sites and reproductive activities.

One of the most definitive examples of age-related variation in responsiveness to host-plant cues does not involve a phytophagous insect at all, but a tachinid parasite of a phytophagous insect. *Eucarcelia rutilla,* a parasite of the pine looper moth, orients to olfactory cues of the looper's host. At the beginning of the preoviposition period, female flies tend to choose oak odor over pine scent in an olfactometer apparatus. As the preoviposition period progresses, the female tachinid fly becomes increasingly responsive to pine odors. By the end of the oviposition period, oak odor is actually repellent to females (Herrebout and van der Veer, 1969). Finally, the data suggest that there is an analogous fluctuation in the contact-discrimination preference for oak or pine as the insect ages. Examples where age has such a pronounced impact on host choice are lacking for phytophagous insects.

Age may often be more precisely specified as the stage in some particular physiological cycle. Host finding by the female cabbage rootfly, *Delia brassicae,* is dependent on the stage in its reproductive cycle. The extent to which host-plant odor increases activity of gravid females is directly proportional to the length of time females have been gravid (Traynier, 1967). In field tests, gravid females are apparently more directionally oriented than those that are not gravid (Hawkes, 1974, 1975; Hawkes and Coaker, 1979), flying upwind in response to odor from cabbage plants.

Searching behavior by ovipositing cabbage butterflies (*Pieris rapae*) is likewise more accurately described with reference to the current fecundity of the individual (Jones, 1977b). Females with large daily complements of eggs left to lay tend to return more frequently to hosts just previously discovered. Butterflies with only a few eggs left to lay quickly leave the vicinity of an encountered host plant. Besides lower directionality, females with more eggs left to lay exhibit higher responsiveness to host plants and shorter average flight length.

Perhaps the most striking relationship between host-location behavior and stage in reproductive cycle is illustrated by the cockchafer, *Melolontha melolontha*. The female cockchafer passes through several ovarian cycles each of which features a cycle in host-location behavior. Reversal in the sense of flight direction over ovarian cycles enables the female to move first to feeding areas and subsequently back to oviposition sites. The cycle in host location has been correlated with ovariole development cycles and neuroendocrinological events (Stengel, 1974). It would be interesting if the sequence of behavior associated with searching for either host type was more carefully examined to determine if it varied not only in terms of habitat selection (via flight direction) but also with respect to the individual's responsiveness to the respective set of host cues. Still, Stengel's work represents the only example in which variation in host use stems from effects of an age-related parameter on host location.

### F. Temperature, Photoperiod, and Other Climatic Factors

Various external climatic factors, notably temperature and photoperiod, affect host-location behavior. Some of these effects are simple and direct. Butterflies, for example, cannot fly until their bodies are above a certain range of temperature (Rawlins, 1980). Below these temperatures, which are themselves a function of ambient temperature, females obviously cannot search for host plants on which to lay their eggs. Furthermore, when direct solar radiation is interrupted momentarily, oviposition rates may markedly decrease (Petersen, 1954; Gossard and Jones, 1977), probably because searching temporarily ceases. Even when temperatures are sufficiently high to permit flight, the frequency of basking periods seems to be greater on cooler days, at least for female pipevine swallowtail butterflies (*Battus philenor*) (Papaj and Rausher, unpublished observation). Although it has not been demonstrated for any herbivorous insect, the time allocated to host-location search may be directly related to the frequency of these thermoregulatory activities that are in turn a function of ambient temperatures.

Other mechanisms for behavioral thermoregulation may induce apparent variation in host-location behavior. If, for example, temperature requirements restrict flight to sunny or shady habitats, then variation in temperature or in responses to temperature may lead to variation in habitat selection which is not directly related to host location. Of course, this variation could influence aspects of host location such as the rates of encounter of alternate host species if those species have different abundances and distributions in different habitats. One must take special care when assessing the ecological significance of habitat selection and host preference to consider these sorts of accessory requirements (Rausher, 1979a).

A more complex way in which temperature impinges on aspects of host-finding behavior and host selectivity is by altering the nutritional state of an insect. Higher temperatures, for example, are associated with higher desiccation rates. Desiccation often stimulates seeking of host plants because these are frequently an important source of water for the animal. Saxena (1969) observed ways in which the orientation of the red cotton bug (*Dysdercus koenigii*) changes as insects become water-stressed. Both high temperatures and water deprivation can decrease the location $T_{50}$ (time in which 50% of the insects found a plant under the conditions of the laboratory assay), apparently by increasing the probability of taxis once an insect enters a "receptive taxis zone." In addition, the number of individuals entering this zone increases as dehydration progresses. Most importantly, dehydrated insects show little discrimination among any of the host-plant species tested. This enhanced orientational response to the succulent leaves of any host type disappears in water-satiated individuals, and host selectivity apparently increases.

Apart from these metabolically derived influences on behavior, temperature and other climatic factors might act as external cues that regulate the development of alternate host-location behaviors. At present, there are no clear examples in which physical cues induce developmental switches of alternate neural pathways controlling host-location behavior. Yet flight polymorphisms in certain herbivorous insects are regulated by climatic factors. In some cases, these polymorphisms are associated with polyethisms in postalighting discrimination and with shifts in host use.

Temperature and photoperiod are among a constellation of external factors that regulate alary polymorphism in aphids and other herbivorous insects (Lees, 1966; Tsitsipis and Mittler, 1976; Hardie, 1980a, 1981). Alary polymorphism, which refers to the occurrence of several morphs differing in their ability to fly, constitutes an example in which variation in behavior is contingent on some variation in the efferent system—in this case, the presence or absence of wings. An apterous (wingless) aphid may possess all of the neural motor patterns for integration of flight and even the motivation to fly, but cannot because it lacks the appropriate mechanical apparatus for flight. The consequence, nevertheless, is a significant contrast in movement patterns by the two forms that profoundly influences host selection. Alate (winged) and apterous forms of aphids and other insects often utilize different species of host plants (for review see Dixon, 1977). Although this observation suggests that the two forms would differ in discriminatory behavior, existing evidence is conflicting. Alate aphids tend to have more sensoria in their antennae (Kennedy *et al.*, 1959a), but the difference is slight (Lees, 1966) and no functional significance has been rigorously attached to it. Indeed, the host-finding abilities of aphids, in general, has been characterized as "degenerate" and "crude" (Kennedy and Fosbrooke, 1973). Many aphids exhibit a rather nonspecific attraction to yellow (Kennedy *et al.*, 1959c), and settling is distributed equally on hosts and nonhosts (Kennedy *et al.*, 1959a,b). Only after landing on a plant can aphids determine if it is a host; host selection by aphids is thus wholly a matter of staying or leaving (Kennedy and Fosbrooke, 1973). Host preferences in these aphid populations are therefore, behaviorally at least, a consequence of postalighting discrimination and of the constraints on dispersal, if any, imposed by alary morphology.

Differences in observed host preferences between winged and wingless generations of aphids does not necessarily reflect variation in postalighting responses to alternate host types. Shifts in host use that accompany seasonal variation in alary morphology are most often mediated by a temporal change in the relative quality of the two host types (e.g., Kennedy and Booth, 1951). Apparently, shifts in relative host quality result in changes in the stimuli relevant to host selection. Thus the aphids themselves may not vary in their response to fixed host types; rather, the plants vary seasonally in the attractiveness of emitted stimuli. Nevertheless, where it has been studied, host alternation that is primarily precipitated by temporal changes in relative host-plant quality (and in the nature of host-selection cues) is at least reinforced by a seasonal change in discrimination behavior of successive generations. The black bean aphid *Aphis fabae* shifts from spindle (*Euonymus europaeus*), on which it overwinters, to sugar beet (*Beta vulgaris*) and beans (*Vicia faba*). The switch occurs when either spindle or sugar beet become mature and less preferred than the young, growing tissue of the alternate host. Hence, host alternation is structured mainly by the pattern of senescence of the alternate hosts. Overlaying this process, however, is a correlated shift in discrimination among particular plant species and among leaves of different age from gynoparous to alate to apterous forms (Kennedy and Booth, 1954). Gynoparae, which colonize young spindle, show the strongest and weakest tendencies to settle on spindle rather than sugar beet and growing rather than mature leaves, respectively. By contrast, apterae, which colonize sugar beet, have the strongest and weakest settling affinities for spindle and growing leaf tissue, respectively. Alatae are intermediate in selectivity. Because there is never a change in the rank order of selectivity, this variation in discrimination is probably not primarily responsible for the observed host shift. Hardie (1980a, 1981), however, has provided evidence that conflicts with this interpretation. Using bean rather than sugar beet as the representative summer host in laboratory assays, he demonstrated that, whereas the gynoparae tend to larviposit more often on spindle than on bean, the alate virginoparae larviposit more frequently on bean than spindle (Hardie, 1980a), when each is reared under the photoperiods normally encountered in the field. Whether host alternation in *A. fabae* is achieved by a shift in discriminatory behavior over successive generations thus remains an open question.

The bird cherry–oat aphid *Rhopalosiphum padi* represents an apparently well-established example in which seasonal host alternation is principally mediated by a shift in discriminatory behavior over successive generations (Dixon, 1971). In postalighting choice tests using an array of potential host types, apterous forms overwhelmingly chose oats, one of several cereals and grasses they normally exploit. Alate gynoparae, by contrast, most frequently selected bird cherry (*Prunus padus*).

In neither of these last two examples is variation in host preference within the aphid populations likely to have arisen from variation in host-finding behavior. All tests, by their small-cage design, restricted the interpretation of results to differences in contact discrimination. In light of the fruitless attempts to demonstrate any host-finding behavior of note in aphids, the limitations of these assays are somewhat forgiveable. Ironically, differences in host-finding behavior between apterous and alate forms may be restricted to the nymphal stages of the aphids, before the wing trait is expressed. According to Hardie (1980b), nymphal apterae of tbe black bean aphid *Aphis fabae* are more likely to move off crowded plants than nymphal alatae. Hardie speculates that the differences in dispersal behavior may ensure that most of the diminishing food supply is left to the alates.

### G. Conspecific Insect Density

The significance of conspecific herbivore density in altering patterns of host use through effects on host-selection behavior is not well known. Yet the presence of conspecifics is a criterion for selection of host plants by some phytophagous insects. Some ovipositing butterflies, for example, avoid laying eggs on plants that bear previsouly laid eggs (Gilbert, 1975; Rothschild and Schoonhoven, 1977; Rausher, 1979b). Egg recognition, though, appears to take place after butterflies have found a plant (Rausher, 1979b), although odor cues perceived at short distances from the plant may be involved as well (Rothschild and Schoonhoven, 1977). Similarly, a marking pheromone that deters oviposition by the apple maggot fly *Rhagoletis pomonella* is probably detected after location of the host site (Prokopy, 1972). In fact, phytophagous insects have apparently not been shown to discriminate among hosts differing in herbivore density prior to arrival on individual host plants.

Almost all evidence implicating conspecific density as a source of variation in host location addresses alary polymorphism in aphids. Crowding on host plants generally triggers the development of alate forms, facilitating dispersal, which eventually leads to discovery of new sites for feeding and reproduction (Lees, 1966, 1967; Dixon and Glen, 1971; Dewar, 1976; Kidd, 1977). Usually, aphids are susceptible to crowding during the nymphal stages, but crowding experienced by the mother can affect the alary morphology of her offspring as well. Even in aphids dominated by winged forms such as the sycamore aphid (*Drepanosiphum platanoides*), crowding as nymphs or freshly molted adults will increase the tendency to disperse before reproducing and hasten the accomplishment of dispersal (Dixon, 1969). Dispersal is apparently represented by both trivial and migratory flight in this insect and, because both types end with the discovery of a new host individual, both pertain to host location. Both prenatal and postnatal crowding of black bean aphids (*Aphis fabae*) induce alate forms (Shaw, 1970a). Furthermore, crowded parents produce a larger percentage of alate migrants that fly before depositing nymphs than uncrowded parents whose progeny are maintained at high densities (Shaw, 1970b). So-called flyers, which deposit their nymphs before flight, act to spread aphid infestations to nearby host plants; migrants spread the infestations over much greater distances. Among alates, there is also a certain proportion of nonflyers, which deposit all of their nymphs without flying. Shaw (1970c) admits that this distinction between flight types is somewhat artificial. Alatae apparently exhibit a so-called continuous polymorphism from nonflyers to migrants that is regulated by crowding levels experienced by parents and nymphs.

The migratory locusts (e.g., *Schistocerca gregaria*) represent one of the best examples of the effect of crowding on host-location behavior. Two phases are common among the locusts, differentiable by a large suite of polymorphic characters. The *gregaria* phase features long tegmina, dark, bold color patterns, larger size at hatch, high fecundity, and a tendency to be gregarious. The *solitaria* phase is characterized by short tegmina, a green, cryptic color pattern, small size at hatch, low fecundity, and no particular inclination to congregate (Kennedy, 1961b). There are also striking polyethisms that have some bearing on host location. The gregarious morph is migratory, manifesting a suppression of response to vegetative stimuli. Whereas solitary individuals evince a visual attraction to upstanding objects such as vegetation and tend to climb and feed periodically on plants, the gregarious form dodges around upstanding objects and is arrested only briefly on plants it encounters (Kennedy, 1975). The transition from solitary locusts to so-called hoppers is primarily directed by crowding of either the mother or the nymph. It may take several generations of continued crowding for the gregaria traits to become fully expressed. Once initiated, alteration of environmental conditions precipitating high density will have little effect on the progression of the migratory form. Gregarious behavior virtually guarantees crowding independent of the environment, and the eventual complete manifestation of the gregarious phase is thus ensured (Kennedy, 1956, 1961b, 1975).

In conclusion, conspecific density does induce variation in host-finding behavior in a variety of herbivorous insects. In theory, avoidance of high densities might moderate host specificity in herbivorous populations by acting to brake accumulations of insects on host types that would be favored at equal densities. Despite the plausibility of this hypothesis, the contention that patterns of host use are altered by effects of conspecific density on host selection is not supported by available evidence.

### H. Maternal Effects

Maternal effects are particularly important to consider when designing experiments that propose to estimate the heritability and selection response of host-location traits. By contributing to resemblances among individuals and to phenotypic differences between families, maternal effects can lead to overestimation of genetic variance. Natural selection can only act on additive genetic variance and cannot act directly through maternal effects, underscoring the need to distinguish between the two (Falconer, 1980). Aside from genetic considerations, maternal effects on host-finding behavior can potentially modify host preferences in phytophagous populations.

The best documented cases of maternal effects on host location in herbivorous insects involve locust migratory and aphid alary polymorphisms. Crowding or certain photoperiod regimes experienced by the aphid mother induces production of alate progeny (Lees, 1966, 1967; Dixon, 1969; Forrest, 1970; Shaw, 1970a; Dixon and Glen, 1971; Dewar, 1976; Hardie, 1981). The migratory locusts represent a classic example of a maternally regulated interplay between migratory activity and host-location behavior (Kennedy, 1975; see also Section II, G). In neither case, however, are maternal effects as likely to structure host preference in a herbivore population as in the gypsy moth *Lymantria dispar*.

Large early-instar *Lymantria dispar* larvae are more likely to repeatedly disperse in laboratory assays than small larvae (Capinera and Barbosa, 1976; Barbosa *et al.*, 1981). The size of hatching larvae is proportional to the size of the egg (Barbosa and Capinera, 1978), and egg size is markedly affected by the quality of the host plant on which the mother was raised as a larva (Barbosa *et al.*, 1981). Females raised on maple produce smaller (and fewer) eggs than those raised on oak (Capinera and Barbosa, 1977; Barbosa *et al.*, 1981). Hence, maternally regulated effects of host type on fecundity. Although it is not yet clear how these factors interact to generate observed patterns of host use, the work by Capinera, Barbosa, and colleagues represents the best available data in which maternal effects have differential impact on the utilization of alternate host types.

A maternally influenced polyethism for host-finding activity with an equally high potential for modifying host preference was reported for the western tent caterpillar *Malacosoma pluviale* (Wellington, 1957, 1960, 1964, 1965). Wellington distinguished several types of larvae issuing from a typical *M. pluviale* egg mass. Type I larvae were very active and moved in a directed fashion toward a light source in a laboratory assay. No Type II larvae, by contrast, showed spontaneous, directed movement in the early instars. Type II larvae diverged further in their level of activity. Type IIa caterpillars were the most active and were observed to follow Type I larvae and their trails. Type IIb individuals were sluggish and consumed less food before pupation than the previous types. Finally, Type IIc larvae were described as extremely sedentary; few reached pupation and those that reached adulthood seldom reproduced.

This disparity in levels of activity and taxis was reflected in the success of the early instars of various types in finding food. Type I individuals easily located food sources (provided in the form of twigs from host trees) by themselves. Type IIa larvae found food easily if a Type I larva was present to guide them. Even without the presence of Type I larvae, they occasionally managed to find food a short distance away. Type IIb and IIc individuals rarely found food from any distance and almost never unless active individuals were present.

These behavioral differences were apparently transmitted to the next generation; active individuals produced a higher proportion of active individuals among their progeny than sluggish individuals (Wellington, 1964). Wellington (1965) also reported tbat most of the active (Types I and IIa) larvae were laid first in an egg mass; most Type IIb and IIc larvae were laid in the last half of an egg mass. He suggested that the progressive decline in progeny quality was due to nutritional depletion. Finally, Wellington (1965) provided data bearing on a maternal effect of diet on the proportion of behavioral types represented among the progeny. In that experiment, Type I larvae were reared on identical diets except that one group was starved during the early instars. Wellington claimed that the group that was deprived of food later produced progeny of which a higher proportion exhibited "reduced vigor" compared to larvae produced by the group that was fed to repletion. If true, this intriguing conclusion provides a basis for speculating that host types that differ nutritionally might vary in herbivore load through maternal effects on host-finding behavior.

Unfortunately, Wellington's analysis (1965) was not statistically rigorous.

Absolute numbers of each type produced by partially starved and well-fed parents were compared invalidly, even though the average egg mass size differed markedly between treatments ($\bar{N}$ for well-fed parents was 157; $\bar{N}$ for food-deprived parents was 78). Furthermore, inviable eggs were treated in the analysis as Type IIc individuals, presumably because they represented the ultimate extreme in sluggish behavior! Finally, and equally important, Wellington performed separate pairwise tests on each behavioral type as though the proportion of one type in an egg mass was independent of the proportion of other types in the egg mass. If Wellington's data are reanalyzed as the percentage of a type among all individuals successfully hatching, a very different pattern emerges (Table I). The differences between the frequencies of activity types issuing from a representative egg mass of well-fed and poorly fed Type I mothers (i.e., an egg mass of a size based on Wellington's means) are not statistically significant; even if they were, the trends are mostly counter to those deduced from Wellington's analysis. In summary, poor nutrition apparently does reduce average egg mass size and percentage hatch of western tent caterpillars, but does not directly influence the proportion of individuals exhibiting certain levels of host-finding activity. Furthermore, any concomitant effect of maternal nutrition on the social structure and success of the tent caterpillar colony is thus the result of variation in group size and is not attributable to changes in the proportions of the behavioral types.

Attempts to replicate demonstration of a polyethism in tent caterpillars has produced somewhat checkered results. Meyers (1978) was unable to discern a relationship between the behavioral type of an individual and its position in the egg mass in the same tent caterpillar species. Greenblatt and Witter (1976) reported a similar polyethism for activity level in the forest tent caterpillar *Malacasoma disstria*. Unfortunately, they too used invalid tests for individual variation in activity level. Greenblatt and Witter classified larvae as active if they traveled 3 cm from a common starting line toward a discrete light source, and sluggish if they did not. (Whether the 3 cm refers to net displacement is unspec-

**TABLE I.** Proportion of Activity Types among Progeny of Well-Fed and Poorly Fed Type I Tent Caterpillar Mothers (*Malacosoma pluviale*)[a,b]

| Nutritional Status of Parents | Activity Types | | | |
|---|---|---|---|---|
| | I | IIa | IIb | IIc |
| Well fed | 0.21 | 0.26 | 0.52 | 0.018 |
| Food deprived | 0.23 | 0.27 | 0.46 | 0.037 |

[a]Data taken from Wellington (1965).
[b]$\chi^2 = 0.8, p > 0.5$.

ified; if it does, it seems plausible that differences in turning rate [i.e., klinokinesis] or phototaxis as well as activity level [i.e., orthokinesis] would lead some individuals and not others across the 3-cm finish line.) Each larva was tested at three different times. Larvae were then grouped according to their composite score into one of eight categories (Table II). Greenblatt and Witter contended that, if all larvae were behaviorally similar and varied randomly over trials with respect to activity level, all eight possible composite scores would be equally represented among all larvae. Because they were not (see observed frequencies in Table II), Greenblatt and Witter concluded that forest tent caterpillars exhibit a polymorphism for activity level. The null hypothesis that larvae will be uniformly distributed throughout all behavioral categories if there was no polyethism, however, is inappropriate. It assumes that larvae that are behaviorally homogeneous will have a probability equal to 0.50 of moving 3 cm in the allotted time when there is no a priori reason for specifying any value for this probability.

As a more appropriate null hypothesis, suppose all larvae are identical in behavior and, under the conditions of the laboratory assay, have an unspecified and estimable probability of completing the "3-cm dash." The best estimate of this probability is taken as the percentage of all larvae that were classified as active; three estimates can be made, one at each time larvae were tested. The probability of being classified as active estimated in this manner averages about 0.23 (and not 0.50 as Greenblatt and Witter assumed), varying somewhat across test periods. The expected frequencies within each of the eight categories are computed by multiplying the appropriate probabilities of being active or sluggish in each test period. An estimate of the total number of larvae in each category is given by the product of these frequencies and the total number of larvae ($N$ =

**TABLE II.** $\chi^2$ Goodness-of-Fit Reanalysis of Forest Tent Caterpillar Polyethism Data[a]

| Behavioral Category[b] | Observed Frequencies | Expected Frequencies | Deviation from Expectation |
|---|---|---|---|
| AAA | 77.35 | 30.47 | + |
| AAS | 98.45 | 75.00 | + |
| ASA | 77.35 | 93.76 | − |
| SAA | 168.77 | 133.61 | + |
| ASS | 182.83 | 234.40 | − |
| SAS | 227.37 | 332.85 | − |
| SSA | 346.90 | 412.54 | − |
| SSS | 1167.30 | 1029.53 | + |
| | 2346.32 | 2342.53 | $\chi^2 = 16.13, p < 0.005$ |

[a]Data from Greenblatt and Witter (1976).
[b]A, active; S, sluggish.

2344). A $\chi^2$ goodness-of-fit test still yields a very significant deviation from the null hypothesis. Inspection of the deviations from expectation (Table II) shows that the extreme classes (AAA and SSS) are larger than expected, suggesting that there is indeed a polyethism. In other words, more larvae are consistently active or sluggish than would be expected by random chance alone. The trend in the intermediate classes, however, is ambiguous. Under the polyethism hypothesis, all intermediate classes should be underrepresented. In fact, four of six classes are smaller than expected according to the null hypothesis, and two are larger.

In summary, despite an invalid statistical test, a woefully arbitrary classification of activity level, and somewhat ambiguous results, Greenblatt and Witter (1976) provide the only statistically suggestive evidence for a polyethism in tent caterpillars. Furthermore, the only report of maternal influence of diet on tent caterpillar behavior is apparently spurious. We suggest the issue be reexamined with the use of ecologically relevant behavioral assays and appropriate statistical analysis, particularly in light of the ramifications on population regulation (Wellington, 1957), evolution of social structure (Ricklefs, 1973), and host preference that have been attached to this work.

## III. GENETIC VARIATION AND NEUROLOGICAL MODELS OF HOST SELECTIVITY

Despite the obvious importance of genetic variability in host-location behavior for understanding the evolution of host races and host diet breadth, for developing biological control techniques and host resistance, and for understanding behavioral mechanisms of host location, only a handful of attempts have been made to assess the contribution of genetic differences to individual variability in behavior. Consequently, the scope for interpretation of the implications of these genetic differences is severely limited. Because other workers have reviewed the ecological and evolutionary implications (e.g., see Futuyma, Chapter 8, this volume), we shall confine our discussion to a description of how genetic investigations can contribute to the development of an integrated model of host-location behavior.

As emphasized earlier, host location involves a sequence of discrimination events beginning with those that initiate movement toward a plant and ending with those that elicit contact. Genetic variation may presumably affect any of the discrimination mechanisms associated with these events. For example, searching behavior may be altered by genetic changes that affect sensitivity to visual or odor stimuli emanating from a host plant. These alterations may affect, among other things, receptor specificity or acceptance thresholds.

Unfortunately, few experiments exist that demonstrate which components of host-finding behavior are influenced by genetic variation. In most studies that report genetic variation in host selection, variation is detected by bioassays of oviposition or feeding preference, in which confined animals are offered two or more host plants, and feeding damage to or numbers of eggs laid on each plant is assessed (Phillips and Barnes, 1975; Hsiao, 1978; Wasserman and Futuyma, 1981; Tabashnik *et al.*, 1981). From such experiments it is usually not possible to determine even whether observed genetic differences in the apportionment of eggs to or feeding time on different hosts are manifested in host-seeking activity or behavior subsequent to host discovery.

Singer (1971, 1983) has provided one of the few examples in which the effects of genetic variation in host preference by phytophagous populations can be localized to particular components of host selectivity by individuals. Populations of the checkerspot butterfly *Euphydryas editha* differ markedly in host use. Most populations are highly monophagous, with primary host species differing among populations, even though in many cases the same or a similar set of host species is available (White and Singer, 1974). Singer has shown that much of this difference in host use is due to differences between populations in postalighting response to host plants. In addition, he has demonstrated that among populations that exhibit the same rank ordering of host species, differences in specificity exist. In other words, in some populations, the second most preferred species becomes acceptable soon after the first, whereas in other populations the intervening discrimination phase is much longer. Because Singer's experiments were done with individuals reared in a common environment, it seems likely that both the differences in ranking and the differences in specificity among populations are due to underlying genetic differences. Preliminary genetic crosses tend to support this inference (M. C. Singer, personal communication).

Investigation of the specific effects of genetic variation on behavioral subcomponents provides a potentially powerful tool for distinguishing between alternate models of the neurological mechanisms underlying host location. As an illustration of the approach, consider the basic model of postalighting discrimination in *Euphydryas editha* proposed by Singer (1982) and Rausher (1983) and described in Section II,B. Host ranking is determined by the rank order of acceptance thresholds, whereas the length of the period during which females discriminate between two hosts (a component of specificity) is determined by the differences in times after oviposition at which those hosts become acceptable (i.e., by the difference in acceptance thresholds).

Depending on the neural mechanism that assigns host species to acceptance thresholds, host ranking and host specificity may be two genetically independent characters or simply different outward manifestations of the same simple discriminatory mechanism. Perhaps the simplest model that can be imagined is one in which there is for each host species a direct and independent neural connection between receptor cells and the motor cells involved in oviposition. Such a situation might be the case, for example, if host plants were discriminated peripherally by receptors specific to secondary compounds that act as stimuli unique to particular host species (Fig. 2). In Model I, the acceptance threshold for a particular host species would be determined by the balance between excitatory and inhibitory inputs to the motor neurons. A separate excitatory pathway would be associated with each host, with the strength of the excitatory input differing for each host. Inhibitory input would be high just after oviposition, then decay over time. Females would exhibit an oviposition response to a particular host plant when the inhibitory input was no longer strong enough to prevent stimulation of the motor neurons by signals from the receptor cells associated with that plant. As indicated by the heavier line in Fig. 2, the connecting cell CII provides a stronger input than cell CI. Consequently, as inhibition falls, excitation will exceed inhibition along pathway II at an earlier time than along pathway I. The only way host ranking may be changed is by changing the strength of the input signals in one or both of the pathways, which means that specificity will also change.

An alternate model (Fig. 2B) is one in which input from receptor cells is received at a central processing center (CPC) where it is assigned to one of several pathways that correspond to acceptance thresholds (pathways corre-

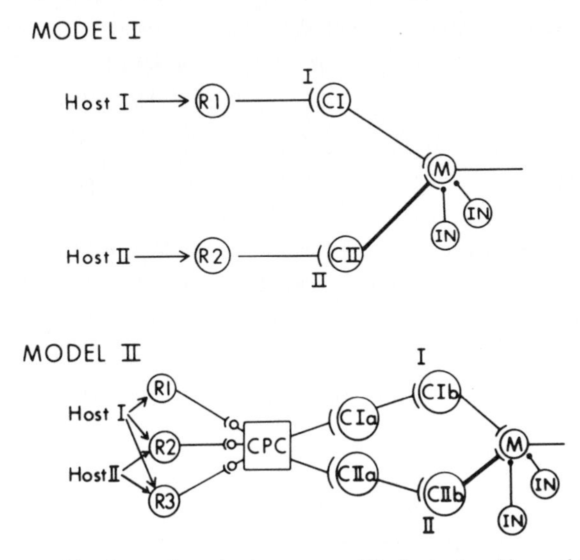


**Fig. 2.** Alternate models of neural mechanisms responsible for host ranking and host specificity. Model I: pathways corresponding to different host species are independent of each other. R1 and R2, receptor cells specific to host species I and II, respectively; CI and CII, connecting neurons; M, motor neuron associated with abdomen-curling and oviposition; IN, inhibitory neurons providing information about acceptance threshold. Width of axons in connecting cells proportional to strength of impulse delivered. Model II: a central processing center (CPC) that has been inserted between receptor (R1–R3) and connecting (CIa, CIb, CIIa, CIIb) cells.

sponding to cells CI and CII in Fig. 2B). Such a central processing center could, for example, be an area where host discrimination occurs by comparing across-fiber patterns of input to templates corresponding to different host species (Schoonhoven and Dethier, 1966; Dethier and Schoonboven, 1969). In this model, host ranking and host specificity may be altered independently: specificity by altering the degree of input delivered by cells CIb and CIIb; and ranking by simply altering which host-plant signal becomes associated with which acceptance pathway.

The contrast between these two models is admittedly somewhat artificial. Neither model should be taken as more than a radical simplification of the actual neurophysiological mechanisms involved. Nonetheless, they do capture the basic elements of two major schools of thought concerning mechanisms of host-plant recognition (Dethier, 1976; Rausher, 1983). Moreover, they serve to illustrate how investigation of genetic variability may provide guidance in choosing between different models of behavioral mechanisms. The second model (Fig. 2B) predicts that host ranking and host specificity should be genetically independent traits; in other words, genetic variants might be found that alter host ranking but not the position of the acceptance thresholds themselves (i.e., specificity). This is possible because genetic variation affecting the central processing center need not affect the strength of input emanating from the acceptance pathways. In effect, acceptance thresholds exist as slots independent of which host plants become plugged into those slots. By contrast, Model I (Fig. 2A) predicts that host ranking and host specificity are not genetically independent, because host ranking can only be altered by genetic changes in input from cells CI and CII, which might also change specificity. By appropriate genetic analysis, it should be possible to determine whether genes affecting host ranking and host specificity reside on the same or different loci, and hence whether Model I or II is more appropriate. The means by which such genetic investigation can be used to supplement neurophysiological investigations of all types of host-location mechanisms would seem to be limited only by the amount of genetic variability that is present (or can be induced) and by the imagination of the investigator (Hoy, 1974, 1978).

## IV. CASE STUDY: INDIVIDUAL VARIATION IN HOST LOCATION BY *BATTUS PHILENOR*

The pipevine swallowtail butterfly, *Battus philenor,* and its associated *Aristolochia* host plants in the Big Thicket of East Texas have proved to be a model system for assessing the sources of variation in host-location behavior by a phytophagous insect. Because the butterfly is very common at certain times of the year and because individual butterflies can be stalked by interested investigators for long periods of time, it has been possible to gather extensive and detailed records of the searching behavior of individual ovipositing female butterflies. Also, for rather serendipitous reasons that will be made clear in the following section, the behavior has been assayed with a great deal less sampling error and fewer arbitrary measures than is usually feasible with field studies on behavior.

### A. Pattern of Individual Variation in Oviposition Behavior

The first pulse of adult female pipevine swallowtail butterflies emerges from pupal diapause at about the time the two host plants, *Aristolochia reticulata* and *A. serpentaria,* begin producing new leaves in the early spring. Both host plants are small, erect perennial herbs that die back at the end of the growing season. Whereas *A. reticulata* produces broad ovate leaves, *A. serpentaria* is characterized by narrow lanceolate leaves. In their relatively short lifetimes (about 2 weeks), female butterflies spend much of their time searching among the vegetation for host plants on which they lay an average of two to three eggs. Searching females periodically alight on the leaves of plants upon which they rapidly tap their foretarsi. In this manner, they apparently taste the plant (Chun and Schoonhoven, 1973; Feeny *et al.*, Chapter 2, this volume). If it is not a host plant, they take flight and resume search. If it is a host plant, they flutter about the plant, frequently alighting. In this way, the female apparently determines the suitability of the host for oviposition (Rausher, 1980; Rausher and Papaj, 1983a). After a variable period of time, the female either oviposits or resumes search.

At any one time, *Battus philenor* females search in one of two modes (Rausher, 1978). Females in the broad-leaf mode alight on more broad leaves than narrow leaves. Females in the narrow-leaf mode alight on more narrow leaves than broad leaves. Furthermore, a searcher in the broad-leaf mode discovers proportionately more of the broad-leaf host plant *A. reticulata,* whereas a searcher in the narrow-leaf mode discovers proportionately more of the narrow-leaf host plant *A. serpentaria* (Rausher, 1978, 1980; D. R. Papaj, unpublished). In both broods, the distribution of search modes is bimodal, that is, individuals may be observed exhibiting either search mode (Fig. 3). Interestingly, most females in the first brood (Brood 1) exhibit a broad-leaf search mode, whereas most Brood 2 females exhibit a narrow-leaf search mode (Fig. 3; Rausher, 1980).

The observed seasonal shift in searching behavior is correlated with a seasonal change in the relative suitability of the two host plants for juvenile survivorship (Rausher, 1980). During Brood 1, both host plants are equally suitable for larval survivorship. Most females are broad-leaf searchers because *A. reticulata* is much more abundant than *A. serpentaria* and, consequently, host-discovery rates are higher. By the time Brood 2 adults emerge, the leaves of *A. reticulata* have

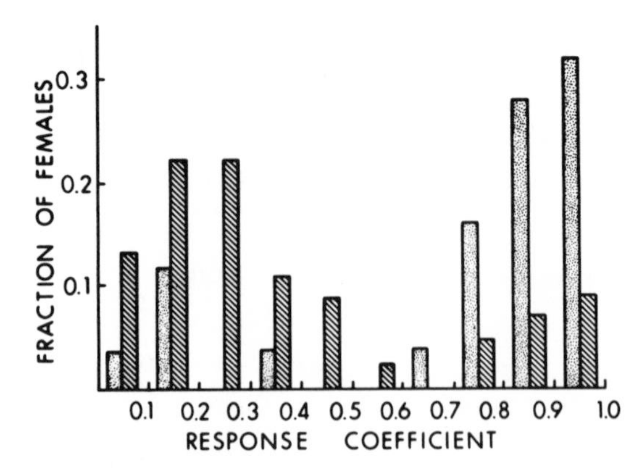


**Fig. 3.** Shift in search modes between Brood 1 and Brood 2. Search modes are estimated by computing response coefficients (RC) for each female. The response coefficient is the fraction of nonhost plant leaves alighted upon during an observation period that are broad. Females with low RCs (0–0.6) are considered narrow-leaf searchers; females with high RCs (0.6–1.0) are considered broad-leaf searchers. Leaf shape of nonhosts is classified as broad or narrow according to length:width (L:W) ratios (see Fig. 4). Dotted bars, RC distribution for Brood 1 females; hatched bars, RC distribution for Brood 2 females. (Data from Rausher, 1980, © Society for the Study of Evolution, 1980; reprinted by permission.)

become sclerophyllous and low in nitrogen, changes that reduce larval growth rates and survivorship relative to *A. serpentaria,* the leaves of which remain suitable for larval growth and survival (Rausher, 1981b). In short, the narrow-leaf search mode of most Brood 2 females ensures that most eggs are laid on the more suitable host species. Whereas the seasonal change in dominant search modes generates, in large part, the observed shift in the apportionment of eggs between the two plants (i.e., the oviposition preference) across broods, the shift is also propagated through seasonal changes in the postalighting responses on the two host plants (Table III). The reduction in the probability of oviposition upon a discovered plant from Brood 1 to Brood 2 is significantly greater when the plant is *A. reticulata* (Rausher, 1980). We note that had behavioral selectivity been

**TABLE III.** Probability of Oviposition by *Battus philenor* on a Plant Once Discovered for Each Host-Plant Species in Broods 1 and 2 (1977)[a]

| Plant Type | Brood 1 | Brood 2 |
|---|---|---|
| *A. reticulata* | 0.23 | 0.12 |
| *A. serpentaria* | 0.64 | 0.47 |

[a]Data from Rausher (1980).

assayed by the small-cage, contact-discrimination tests prevalent in studies of host preference, the data would likely have had the form of those given in Table III. Specifically, an apparent preference for *A. serpentaria* would have been inferred when, in fact, most Brood 1 females are searching for and laying most of their eggs on *A. reticulata*. Furthermore, the cage-assayed selectivity would have greatly underestimated the degree of oviposition preference for *A. serpentaria* observed in the second brood because most Brood 2 females never or very rarely alight on *A. reticulata*. This anomaly underscores the fact that behavioral selectivity measured in terms of prealighting and postalighting behavior will not necessarily be equivalent in sign or magnitude.

Before the sources of individual variation in searching behavior could be determined, it was necessary to show that the differences in search modes were not an artifact of variation in the vegetation structure over which females were observed. In particular, it was imperative to demonstrate that the bimodal distribution of search modes was not the result of patchiness in space or time. Were narrow-leaf searchers, for example, those that happened to be flying through grassy areas when they were observed? In the first brood of 1981, we were able to examine this hypothesis by sampling the relative density of broad and narrow leaf shapes along the flight paths of searching females whose search modes were simultaneously determined (Rausher and Papaj, 1983b). Figures 4 and 5 show one result of that investigation. First, it will be noted (Fig. 4) that the leaf-shape distribution is conveniently bimodal, thus alleviating an otherwise arbitrary distinction between broad and narrow leaf shapes. Second, the proportion of broad and narrow leaves in quadrats sampled along the flight paths of searching females are obviously quite similar for females with broad, narrow, or intermediate search modes (Fig. 5).

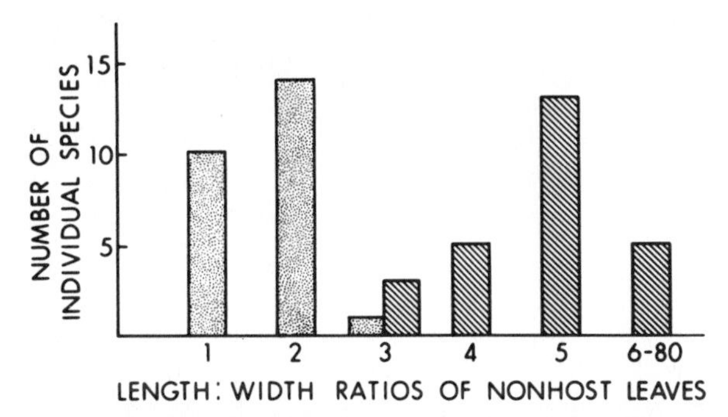


**Fig. 4.** Frequency distribution of nonhost plant species in study area categorized by leaf L:W ratio and their corresponding broad and narrow leaf-shape assignment. Dotted bars, species designated as broad-leaved; hatched bars, species designated as narrow-leaved. For comparison, the L:W ratio for *A. reticulata* is 1.63; for *A. serpentaria*, 15.02. (From Rausher and Papaj, *in press*. © Baillere Tindall, 1982; reprinted by permission.)

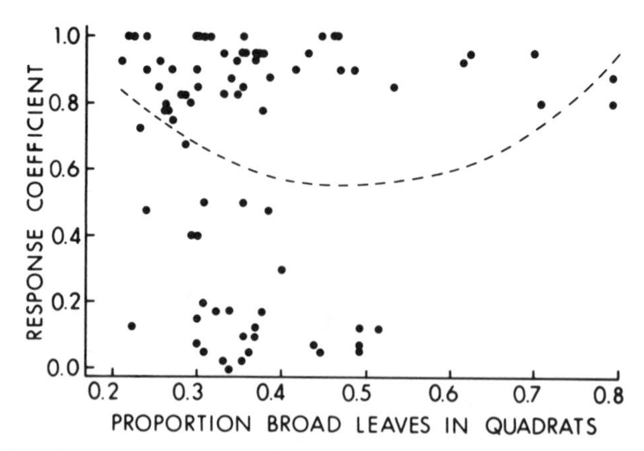


**Fig. 5.** Relationship between response coefficient (RC) and proportion of nonhost leaves in flight-path quadrats that were broad (PB). Best-fitting quadratic regression (dashed line) is not significant. (From Rausher and Papaj, *in press*. © Baillere Tindall, 1982; reprinted by permission.)

Thus the observed individual variation in search modes does indeed reflect a real difference in response to leaf shape and is not an artifact of variation in conditions under which females were observed. In particular, differences in search modes among individuals cannot be explained by temporal or spatial heterogeneity in leaf-shape distributions.

## B. Possible Sources of Variation in Searching Behavior

Three hypotheses suggested in the previous section can account for the observed shift in the predomenant leaf-shape search mode of ovipositing *Battus philenor* butterflies: genotypic differences between early- and late-spring individuals; a developmental switch in adult searching behavior regulated by environmental cues such as temperature and photoperiod to which the preimaginal stages are exposed; or behavioral plasticity directed and focused by the experience of the searching adult. Each of these hypotheses is briefly described here.

### *1. Genotypic Differences*

The genotype of the individual may, in large part, determine its particular leaf-shape search mode. Differences in genotypic frequencies between the broods could be maintained by correlative differences in the tendency to enter pupal diapause. In other words, genes conferring a tendency to exhibit a narrow-leaf search mode may be linked with genes conferring a tendency not to diapause under early-spring conditions. In this manner, Brood 1 progeny with broad-leaf search genes would emerge during Brood 1 of the following year, and Brood 1 progeny with narrow-leaf search genes would emerge during Brood 2. This diapause pattern is consistent with what little is known about pupal diapause in East Texas populations of *Battus philenor* (Rausher, 1979c). Linkage disequilibria between the two gene complexes could theoretically be maintained by any one of several mechanisms: (1) formation of a supergene incorporating both gene complexes and acting as a single unit of recombination; (2) inversions containing both gene complexes and acting as a single selection unit; (3) strong selection against nondiapausing broad-leaf searchers and diapausing narrow-leaf searchers; or (4) assortative mating (Ford, 1964). Alternatively, a heteroallelic switch gene controlling both diapause and search gene complexes (which themselves are not necessarily linked and for which all members of the population are functionally homozygous) may regulate the correlated expression of the traits. Genetic polymorphisms in diapause length like those postulated have been reported in other insects (Waldbauer and Sternburg, 1973; Waldbauer, 1978). Correlations between diapause tendency and other traits are known for other insects (Istock *et al.*, 1976) as well.

### 2. *Environmentally Induced Developmental Switch*

The physical environment under which a female develops in the preimaginal stages could influence ber leaf-shape search mode. If preliminary diapause data (Rausher, 1979c) are correct, Brood 1 adults are developing as larvae and pupae through spring, summer, fall, and winter, whereas Brood 2 adults are developing as larvae and pupae through the spring. Associated differences in environmental conditions such as photoperiod and temperature could cause the observed seasonal variation in dominant search mode. With respect to developmental switches, gerrid alary polymorphism is especially interesting because it arises via a developmental mechanism very similar to the one proposed here. Like *Battus philenor,* certain populations of *Gerris odontogaster* are partially bivoltine. The larger first generation comprises mostly short-winged adults (micropters), whereas the second generation is made up of long-winged adults (macropters) that can fly. Exposure of preimaginal stages to long days that are gradually increasing produces micropters (Vespalainen, 1971); temperature is also thought to play a role in determining wing morphology and flight behavior in natural *Gerris* populations (Vespalainen, 1978). The pattern exhibited by *G. odontogaster* suggests that a developmental switch mechanism is a viable hypothesis to account for the observed seasonal shift in host-location behavior by *B. philenor.*

### 3. *Behavioral Plasticity Regulated by Experience*

For any emerging female, there may not be a priori tendency to exhibit one or the other leaf-shape mode. Rather, the leaf-shape search mode adopted may be the result of experience with the set of host plants available at emergence. The most cogent mechanism leading to such behavioral plasticity is learning. The leaf-shape search mode adopted by a female might be that associated with the host plant on which the oviposition rate is highest. The oviposition rate is a function of both the host-plant discovery rate and the probability of oviposition on a discovered plant. In the first brood when the probability of oviposition given discovery is similar for both plants (Table III), the dominant leaf-shape search mode would be determined mostly by the discovery rates of the alternate hosts. Because *A. reticulata* is much more abundant than *A. serpentaria,* most females adopt a broad-leaf search mode. In the second brood when probability of oviposition on *A. serpentaria* is much greater than on *A. reticulata* (Table III), most females would adopt a narrow-leaf search mode; if this is true, the difference in postalighting probability of oviposition must more than compensate for the difference in discovery rates and must produce a greater initial oviposition rate on *A. serpentaria*. Behavioral plasticity should also be characterized by switching of an individual's search mode when oviposition rates shift to favor the alternate search mode. When the difference between oviposition rates is stable, search modes may become more extreme as the individual gains experience.

### C. Experimental Evidence for the Effect of Experience on Search Mode

A preliminary experiment (D. R. Papaj, unpublished) has helped to distinguish among these three hypotheses. A large enclosure (50 × 50 × 10 ft) was built at the Zoology Field Station in Duke Forest about 2 miles from the Duke University campus in Durham, North Carolina. Inside this enclosure, butterflies will mate, feed at artificial nectar sources, and search for and lay eggs on host plants.

Butterflies collected from a Virginia montane population were reared under constant controlled conditions in growth chambers at Duke. Groups of butterflies were subsequently exposed to plant arrays containing equal numbers of broad-leaved and narrow-leaved nonhost plants, and a cohort of either one or the other host plant, but never both simultaneously. The leaf-shape search mode was assayed in the usual way. (Rausher, 1978; Fig. 3). The results for one experiment (Fig. 6) are striking. Butterflies exposed to *A. reticulata* adopt a broad-leaf search mode; butterflies exposed to *A. serpentaria* adopt a narrow-leaf search mode. Because both groups are presumably genetically similar and were raised under identical developmental regimes, the behavior appears to be very plastic and subject to modification by experience with the host plants. This experiment in which individual groups were exposed successively to *A. reticulata* and then *A. serpentaria* is crudely analogous to the seasonal shift in relative host suitability in East Texas previously described. Future experiments more closely mimicking natural changes as well as field mark–recapture experiments using stocks collected from both Texas broods are planned. These studies may tell us more about the relative importance of genotypic differences, developmental

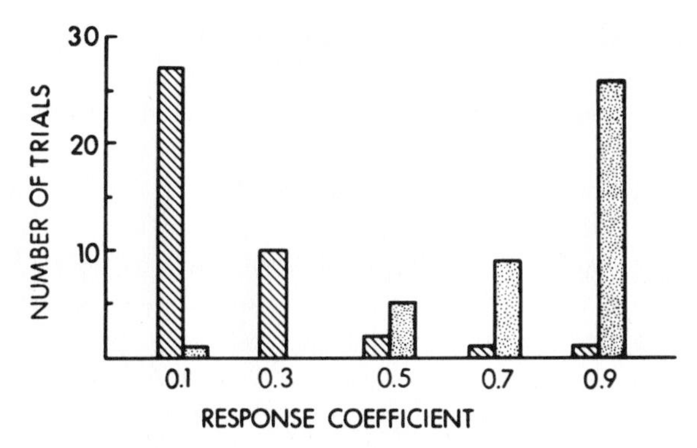


**Fig. 6.** Difference in response coefficients between females exposed to *A. reticulata* and those exposed to *A. serpentaria*. Dotted bars, females exposed to *A. reticulata;* hatched bars, females exposed to *A. serpentaria*. Difference is highly significant ($p = 0.0005$, Kolmogorov-Smirnov two-sample test).

influences, and behavioral plasticity in the seasonal shift of dominant search mode, but the qualitative result is expected to remain the same; learning by females is sufficient to explain the observed seasonal and individual variation in host-finding behavior and resulting oviposition preference.

Corroborative field evidence for the plasticity of search modes was obtained in recent mark–recapture experiments on wild butterflies. Females followed in the field, marked, and refollowed after at least one intervening night, tend to become more extreme in their bias toward alighting on one or the other leaf shape (D. R. Papaj, unpublished). Such a change in search mode would not be expected under either the genetic variation or the developmental switch hypothesis. This pattern of change in performance is also in contrast with findings for *Colias* butterflies, which seem to forget what they have learned in previous oviposition bouts (M. L. Stanton, unpublished). Finally, switching from one mode to another has been recognized in a significant portion of our observations to date; roughly 5% of followed individuals underwent a dramatic shift in search modes either within an observation period or between observation periods (D. R. Papaj, unpublished).

The advantage of being able to switch readily from one search mode to another and the benefit of behavioral plasticity in general probably lies in the ability of an individual to track short-term changes in the relative suitability of alternate host plants. Some preliminary evidence in support of this speculation suggests that the frequency of switches is indeed higher during the time of season when the suitability of *A. reticulata* has declined to the point where oviposition rates on *A. serpentaria* should begin to exceed those on *A. reticulata*. Presumably, those individuals capable of changing their search mode as ecological conditions change to favor the alternate host will have a higher fitness. This supposition remains to be directly tested.

In summary, *Barrus philenor* females employ a type of shape discrimination in host location. Adoption of either a broad or a narrow leaf-shape search mode is correlated with enhanced rates of discovery of the *Aristolochia* host with that leaf shape. The search mode of an ovipositing female is subject to modification according to the set of host plants to which the female is exposed. Search modes are both reversible on occasion and subject to improvement with age and/or experience. Searching individuals thus vary markedly in their responses to leaf shapes encountered. Furthermore, seasonal variation in the dominant search mode exhibited by ovipositing females is primarily responsible for the seasonal shift in the oviposition preference of the pipevine swallowtail butterfly population in east Texas. Although originally described as a search image (Rausher, 1978), data on the ontogeny of leaf-shape search modes presently do not distinguish between conditioning of response and conditioning of perception. Nevertheless, the observation that high densities of vegetation mask host plants and reduce their susceptibility to discovery by ovipositing females suggests that host plants are cryptic in nature (Rausher, 1981a). If this crypticity is based on the confounding of visual cues, search image formation for leaf shape may well be the mechanism of choice of the pipevine swallowtail butterfly.

## V. SUMMARY

The significance of host-location behavior in determining patterns of host use by phytophagous insects has been understated in the past. Nevertheless, a complete understanding of the evolution of host preferences in herbivore populations will require closer examination of the host-finding component of the host-selection behavioral sequence. Abundant evidence suggests that individual variation in traits associated with host seeking contributes to observed differences in host use between or within phytophagous insect populations. Because only genetically based variability can serve as raw material for natural selection to modify host use, it is particularly important to distinguish between genetic and environmental sources of variation in host-location traits pertaining to host selectivity.

Examples of environmental sources of individual variation in host location are common. Some environmental effects on host finding, notably conditioning to or contact with the host plant, are known to influence host preferences markedly. Other sources of variability, such as time-dependent responsiveness and climatic factors, have been shown to alter host use through effects on discrimination subsequent to host discovery and are likely to act through variability in host-seeking behavior as well.

Studies on genetic sources of variation in host-seeking behavior are virtually nonexistent. The outcome of recent work on genetic variation in other aspects of host selection suggests that such variability may also be present in host-location traits. A model is presented that makes use of genetic variation in host-selection behavior to gain insight into the neurophysiology of host selectivity.

Finally, the *Battus philenor–Aristolochia* system provides a case study on individual variation in host location. Ovipositing female pipevine swallowtail butterflies exhibit individual variation in searching behavior that leads directly to variation in host use. This behavioral variability can be attributed largely to the effect of experience with host plants on alighting responses to leaf-shape cues. Though genetic variation may make a contribution, this environmental source of variability is sufficient to account for both individual differences and seasonal changes in searching behavior.

## Acknowledgments

Comments and criticism by William Conner, Vincent Dethier, Haruhiko Itagaki, Maureen Stanton, Sara Via, and Reginald Webster greatly improved the manuscript. We also thank Robert Gendron, Peter Kareiva, Michael Singer, and John Staddon for discussions during which many ideas expressed here were formulated. The senior author is particularly indebted to Rachel Fink for assistance in all phases of preparation of the manuscript.

## REFERENCES

Alcock, J. (1979). "Animal Behavior: An Evolutionary Approach." Sinauer, Sunderland, Mass.

Alloway, T. M. (1973). Learning in insects except Apoidea. *In* "Invertebrate Learning" (W. C. Corning, J. A. Dyal, and A. O. D. Willows, eds.), Vol. 1, pp. 131–171. Plenum, New York.

Atema, J., Holland, K., and Ikehara, W. (1980). Olfactory responses of yellowfin tuna (*Thunnus albacares*) to prey odors: Chemical search images. *J. Chem. Ecol.* **6,** 457–464.

Banks, C. J. (1957). The behaviour of individual coccinellid larvae on plants. *Br. J. Anim. Behav.* **5,** 12–24.

Barbosa, P., and Capinera, J. L. (1978). Population quality, dispersal and numerical change in the gypsy moth, *Lymantria dispar* (L.). *Oecologia* **36,** 203–209.

Barbosa, P., Cranshaw, W., and Greenblatt, J. A. (1981). Influence of food quantity and quality on polymorphic dispersal behaviors in the gypsy moth, *Lymantria dispar*. *Can. J. Zool.* **59,** 293–296.

Barbosa, P., Greenblatt, J., Withers, W., Cranshaw, W., and Harrington, E. A. (1979). Host-plant preferences and their induction in larvae of the gypsy moth, *Lymantria dispar*. *Entomol. Exp. Appl.* **26,** 180–188.

Barton Browne, L. (1975). Regulatory mechanisms in insect feeding. *Adv. Insect Physiol.* **11,** 1–116.

Bernays, E., and Chapman, R. F. (1974). The regulation of food intake by acridids. *In* "Experimental Analysis of Insect Behaviour" (L. Barton Browne, ed.), pp. 48–59. Springer–Verlag, New York.

Capinera, J. L., and Barbosa, P. (1976). Dispersal of first-instar gypsy moth larvae in relation to population quality. *Oecologia* **26,** 53–64.

Capinera, J. L., and Barbosa, P. (1977). Influence of natural diets and larval density on gypsy moth, *Lymantria dispar* (Lepidoptera: Orgyiidae), egg mass characteristics. *Can. Entomol.* **109,** 1313–1318.

Cassidy, M. D. (1978). Development of an induced food plant preference in the Indian stick insect, *Carausius morosus*. *Entomol. Exp. Appl.* **24,** 87–93.

Chandler, A. E. F. (1969). Locomotory behaviour of first instar larvae of aphidophagous Syrphidae (Diptera) after contact with aphids. *Anim. Behav.* **17,** 673–678.

Chew, F. S. (1975). Coevolution of pierid butterflies and their cruciferous foodplants. I. The relative quality of available resources. *Oecologia* **20,** 117–128.

Chew, F. S. (1977). Coevolution of pierid butterflies and their cruciferous foodplants. II. The distribution of eggs on potential foodplants. *Evolution* **31,** 568–579.

Chun, M. W., and Schoonhoven L. M. (1973). Tarsal contact chemosensory hairs on the large white butterfly *Pieris brassicae* and their possible role in oviposition behavior. *Entomol. Exp. Appl.* **16,** 343–357.

Claridge, M. F., and Wilson, M. R. (1978). Seasonal changes and alternation of food plant preferences in oligophagous mesophyll-feeding leafhoppers. *Oecologia* **37,** 247–255.

Copp, N. H., and Davenport, D. (1978). *Agraulis* and *Passiflora*. I. Control of specificity. *Biol. Bull.* **155,** 98–112.

Croze, H. (1970). Searching image in carrion crows. *Z. Tierpsychol. Beih.* **5,** 1–86.

Dawkins, M. (1971a). Perceptual changes in chicks: Another look at the "search image" concept. *Anim. Behav.* **19,** 566–574.

Dawkins, M. (1971b). Shifts of "attention" in chicks during feeding. *Anim. Behav.* **19,** 575–582.

Dethier, V. G. (1954). Evolution of feeding preferences in phytophagous insects. *Evolution* **8,** 33–54.

Dethier, V. G. (1957). Communication by insects: Physiology of dancing. *Science* **125,** 331–336.

Dethier, V. G. (1976). "The Hungry Fly: A Physiological Study of the Behavior Associated with Feeding." Harvard Univ. Press, Cambridge, Mass.

Dethier, V. G. (1978). Studies on insect/host plant relations—past and future. *Entomol. Exp. Appl.* **24,** 759–766.

Dethier, V. G., and Schoonhoven, L. M. (1969). Olfactory coding by lepidopterous larvae. *Entomol. Exp. Appl.* **12,** 535–543.

Dewar, A. M. (1976). The effect of crowding on alate production and weight of apterous exules of the apple-grass aphid, *Rhopalosiphum insertum*. *Ann. Appl. Biol.* **82,** 203–208.

deWilde, J., Bongers, W., and Schooneveld, H. (1969). Effects of hostplant age on phytophagous insects. *Entomol. Exp. Appl.* **12,** 714–720.

Dingle, H. (1965). The relationship between age and flight activity in the milkweed bug, *Oncopeltus*. *J. Exp. Biol.* **42,** 269–283.

Dingle, H. (1968). The influence of environment and heredity on flight activity in the milkweed bug *Oncopeltus*. *J. Exp. Biol.* **48,** 175–184.

Dingle, H. (1972). Migration strategies in insects. *Science* **175,** 1327–1334.

Dingle, H., and Arora, G. (1973). Experimental studies of migration in bugs of the genus *Dysdercus*. *Oecologia* **12,** 119–140.

Dixon, A. F. G. (1959). An experimental study of the searching behavior of the predatory coccinellid beetle *Adalia decempunctata* (L). *J. Anim. Ecol.* **28,** 259–281.

Dixon, A. F. G. (1969). Population dynamics of the sycamore aphid *Drepanosiphum platanoides* (Schr.) (Hemiptera: Aphididae): Migratory and trivial flight activity. *J. Anim. Ecol.* **38,** 585–606.

Dixon, A. F. G. (1971). The life-cycle and host preferences of the bird cherry-oat aphid *Rhapalosiphum padi* L. and their bearing on theories of host-alternation in aphids. *Ann. Appl. Biol.* **68,** 135–147.

Dixon, A. F. G. (1977). Aphid ecology: Life cycles, polymorphism, and population regulation. *Annu. Rev. Ecol. Syst.* **8,** 329–353.

Dixon, A. F. G., and Glen, D. M. (1971). Morph determination in the bird cherry-oat aphid *Rhapalosiphum padi* L. *Ann. Appl. Biol.* **68,** 11–21.

Douwes, P. (1968). Host-selection and host-finding in egg-laying female *Cidaria albulata* L. (Lepidoptera: Geometridae). *Opusc. Entomol.* **33,** 233–279.

Dunlap-Pianka, H., Boggs, C. L., and Gilbert, L. E. (1977). Ovarian dynamics in Heliconiine butterflies: Programmed senescence versus eternal youth. *Science* **197,** 487–490.

Ehrlich, P. R., and Gilbert, L. E. (1973). Population structure and dynamics of the tropical butterfly *Heliconius ethilla. Biotropica* **5,** 69–82.

Falconer, D. S. (1980). "Introduction to Quantitative Genetics." Longmans, Green, New York.

Ford, E. B. (1964). "Ecological Genetics." Wiley, New York.

Forrest, J. M. S. (1970). The effect of maternal and larval experience on morph determination in *Dysaphis devecta. J. Insect Physiol.* **16,** 2281–2292.

Fox, L. R., and Morrow, P. A. (1981). Specialization: species property or local pbenomenon? *Science* **211,** 887–893.

Frankie, G. W., Opler, P. A., and Bawa, K. S. (1976). Foraging behavior of solitary bees: implications for outcrossing of a neotropical forest species. *J. Ecol.* **64,** 1049–1057.

Gendron, R. P., and Staddon, J. E. R. Searching for cryptic prey: the effect of search speed. *Am. Nat.* in press.

Gilbert, L. E. (1975). Ecological consequences of a coevolved mutualism between butterflies and plants. *In* "Coevolution of Animals and Plants" (L. E. Gilbert, and P. H. Raven, eds.), pp. 210–240. Univ. of Texas Press, Austin.

Gossard, T. W., and Jones, R. E. (1977). The effects of age and weather on egg-laying in *Pieris rapae* L. *J. Appl. Ecol.* **14,** 65–71.

Greenblatt, J. A., and Witter, J. A. (1976). Behavioural studies on *Malacosoma disstria* (Lepidoptera: Lasiocampidae). *Can. Entomol.* **108,** 1225–1228.

Hanson, F. E., and Dethier, V. G. (1973). Role of gustation and olfaction in food plant discrimination in the tobacco hornworm, *Manduca sexta* (Lepidoptera: Sphingidae). *J. Insect Physiol.* **19,** 1019–1034.

Hardie, J. (1980a). Reproductive, morphological, and behavioral affinities between the alate gynopara and virginopara of aphid, *Aphis fabae. Physiol. Entomol.* **5,** 385–396.

Hardie, J. (1980b). Behavioural differences between alate and apterous larvae of the black bean aphid, *Aphis fabae:* Dispersal from the host plant. *Entomol. Exp. Appl.* **28,** 338.

Hardie, J. (1981). Juvenile hormone and photoperiodically controlled polymorphism in *Aphis fabae:* Prenatal effects on presumptive oviparae. *J. Insect Physiol.* **27,** 257–265.

Haskell, P. T., Paskin, M. W. J., and Moorhouse, J. E. (1962). Laboratory observations on factors affecting the movements of hoppers of the desert locust. *J. Insect Physiol.* **8,** 53–78.

Hawkes, C. (1974). Dispersal of the adult cabbage root fly [*Erioischia brassicae* (Bouche)] in relation to a cabbage crop. *J. Appl. Ecol.* **11,** 83–93.

Hawkes, C. (1975). Physiological condition of adult cabbage root fly [*Erioischia brassicae* (Bouche)] attracted to host-plants. *J. Appl. Ecol.* **12,** 497–506.

Hawkes, C., and Coaker, T. H. (1979). Factors affecting the behavioural responses of the adult cabbage root fly, *Delia brassicae,* to host plant odour. *Entomol. Exp. Appl.* **25,** 45–58.

Heinrich, B., Mudge, P., and Deringus, P. (1977). A laboratory analysis of flower constancy in foraging bumblebees: *Bombus ternarius* and *B. terricola. Behav. Ecol. Sociobiol.* **2,** 247–266.

Herrebout, W. M., and van der Veer, J. (1969). Habitat selection in *Eucarcelia rutilla* Vill (Diptera: Tachinidae). III. Preliminary results of olfactometer experiments with females of known age. *Z. Angew. Entomol.* **64,** 55–61.

Hillyer, R. J., and Thorsteinson, A. J. (1969). The influence of the host plant or males on ovarian development or oviposition in the diamondback moth *Plutella maculipennis* (Curt.). *Can. J. Zool.* **47,** 805–816.

Hinde, R. A. (1970). "Animal Behavior: A Synthesis of Ethology and Comparative Psychology," 2nd ed. McGraw–Hill, New York.

Hinton, H. E. (1981). "Biology of Insect Eggs." (3 vols.) Pergamon, New York.

Hopkins, A. D. (1917). A discussion of C. G. Hewitt's paper on "Insect Behaviour". *J. Econ. Entomol.* **10,** 92–93.

Hovanitz, W. (1969). Inherited and/or conditioned changes in host-plant preference in *Pieris. Entomol. Exp. Appl.* **12,** 729–735.

Hovanitz, W., and Chang, V. C. S. (1963). Ovipositional preference tests with *Pieris. J. Res. Lepid.* **2,** 185–200.

Hoy, R. R. (1974). Genetic control of acoustic behavior in crickets. *Am. Zool.* **14,** 1067–1080.

Hoy, R. R. (1978). Acoustic communication in crickets: a model system for the study of feature detection. *Fed. Proc. Fed. Am. Soc. Exp. Biol.* **37,** 2316–2323.

Hsiao, T. H. (1978). Host plant adaptations among geographic populations of the potato beetle. *Entomol. Exp. Appl.* **24,** 437–447.

Ichinosé, T., and Honda, H. (1978). Ovipositional behavior of *Papilio protenor demetrius* Cramer and the factors involved in its host plants. *Appl. Entomol. Zool.* **13,** 103–114.

Istock, C. A., Zisfein, J., and Vavra, K. J. (1976). Ecology and evolution of the pitcher-plant mosquito. II. Substructure of fitness. *Evolution* **30,** 535–547.

Ives, P. M. (1978). How discriminating are cabbage butterflies? *Aust. J. Ecol.* **3,** 261–276.

Janzen, D. (1971). Euglossine bees as long-distance pollinators of tropical plants. *Science* **171,** 203–205.

Jermy, T., Hanson, F. E., and Dethier, V. G. (1968). Induction of specific food preference in lepidopterous larvae. *Entomol. Exp. Appl.* **11,** 211–230.

Jones, R. E. (1977a). Search behavior: a study of three caterpillar species. *Behaviour* **60,** 237–259.

Jones, R. E. (1977b). Movement patterns and egg distributions in cabbage butterflies. *J. Anim. Ecol.* **46,** 195–212.

Jones, R. E., and P. M. Ives. (1979). The adaptiveness of searching and host selection behaviour in *Pieris rapae* (L.). *Aust. J. Ecol.* **4,** 75–86.

Kaddou, I. (1960). The feeding behaviour of *Hippodamia quinquesignata* (Kirby) larvae. *Univ. Calif. Publ. Entomol.* **16,** 181–232.

Kennedy, J. S. (1956). Phase transformation in locust biology. *Biol. Rev. Cambridge Philos. Soc.* **31,** 349–370.

Kennedy, J. S. (1961a). A turning point in the study of insect migration. *Nature (London)* **189,** 785–791.

Kennedy, J. S. (1961b). Continuous polymorphism in locusts. *Symp. R. Entomol. Soc. London* **1,** 80–90.

Kennedy, J. S. (1965a). Mechanisms of host plant selection. *Ann. Appl. Biol.* **56,** 317–322.

Kennedy, J. S. (1965b). Coordination of successive activities in an aphid: Reciprical effects of settling on flight. *J. Exp. Biol.* **43,** 489–509.

Kennedy, J. S. (1966). The balance between antagonistic induction and depression of flight activity in *Aphis fabae* Scopoli. *J. Exp. Biol.* **45,** 215–228.

Kennedy, J. S. (1975). Insect dispersal. *In* "Insects, Science, and Society" (D. Pimentel, ed.), pp. 103–119. Academic Press, New York.

Kennedy, J. S., and Booth, C. O. (1951). Host alternation in *Aphis fabae* Scop. I. Feeding preferences and fecundity in relation to the age and kind of leaves. *Ann. Appl. Biol.* **38,** 25–64.

Kennedy, J. S., and Booth, C. O. (1954). Host alternation in *Aphis fabae* Scop. II. Changes in the aphids. *Ann. Appl. Biol.* **41,** 88–106.

Kennedy, J. S., and Booth, C. O. (1963). Coordination of successive activities in an aphid: The effect of flight on the settling responses. *J. Exp. Biol.* **40,** 351–369.

Kennedy, J. S., and Booth, C. O. (1964). Coordination of successive activities in an aphid: Depression of settling after flight. *J. Exp. Biol.* **41,** 805–824.

Kennedy, J. S., and Fosbrooke, I. H. M. (1973). The plant in the life of an aphid. *Symp. R. Entomol. Soc. London* **6,** 129–140.

Kennedy, J. S., and Ludlow, A. R. (1974). Coordination of two kinds of flight activity in an aphid. *J. Exp. Biol.* **61,** 173–196.

Kennedy, J. S., Booth, C. O., and Kershaw, W. J. S. (1959a). Host-finding by aphids in the field. I. Gynoparae of *Myzus persicae* (Sulzer). *Ann. Appl. Biol.* **47,** 410–423.

Kennedy, J. S., Booth, C. O., and Kershaw, W. J. S. (1959b). Host-finding by aphids in the field. II. *Aphis fabae* Scop. (gynoparae) and *Brevicoryne brassicae* L. with a re-appraisal of the role of host-finding behaviour in virus spread. *Ann. Appl. Biol* **47,** 424–444.

Kennedy, J. S., Booth, C. O., and Kershaw, W. J. S. (1959c). Host-finding by aphids in the field. III. Visual attraction. *Ann. Appl. Biol.* **49,** 1–21.

Kidd, N. A. C. (1977). The influence of population density on the flight behaviour of the lime aphid, *Eucallipterus tiliae. Entomol. Exp. Appl.* **22,** 251–261.

Krebs, J. R. (1973). Behavioral aspects of predation. *In* ''Perspectives in Ethology'' (P. P. G. Bateson, and P. H. Klopfer, eds.), Vol. 1, pp. 73–109. Plenum, New York.

Labine, P. A. (1968). The population biology of the butterfly, *Euphydryas editha*. VIII. Oviposition and its relation to patterns of oviposition in other butterflies. *Evolution* **22,** 799–805.

Latheef, M. A., and Irwin, R. D. (1979). Factors affecting oviposition of *Pieris rapae* on cabbage. *Environ. Entomol.* **8,** 606–609.

Lees, A. D. (1966). The control of polymorphism in aphids. *Adv. Insect Physiol.* **3,** 207–277.

Lees, A. D. (1967). The production of apterous and alate forms in the aphid *Megoura viciae* (Buckt) with special reference to the role of crowding. *J. Insect Physiol.* **13,** 289–318.

Menzel, R. (1967). Utersuchungen zum Erlernen von Spektralfarben durch die Honigbiene (*Apis mellifera*). *Z. Vgl. Physiol.* **56,** 22–62.

Menzel, R. (1968). Das Gadächtnis der Honigbiene für Spektralfarben. I. Kurzzeitiges und Langzeitiges Behalten. *Z. Vgl. Physiol.* **60,** 82–102.

Menzel, R., and Erber, J. (1972). The influence of the quantity of reward on the learning performance in honeybees. *Behaviour* **41,** 27–42.

Menzel, R., and Erber, J. (1978). Learning and memory in bees. *Sci. Am.* **239,** 80–87.

Menzel, R., Erber, J., and Masuhr, T. (1974). Learning and memory in the honeybee. *In* ''Experimental Analysis of Insect Behaviour'' (L. Barton Browne, ed.), pp. 195–217. Springer–Verlag, New York.

Meyers, J. H. (1978). A search for behavioral variation in first and last laid eggs of western tent caterpillars and an attempt to prevent a major population decline. *Can. J. Zool.* **56,** 2359–2363.

Moorhouse, J. E. (1971). Experimental analysis of the locomotor behaviour of *Schistocerca gregaria* induced by odour. *J. Insect Physiol.* **17,** 913–920.

Murton, R. K. (1971). The significance of a specific search image in the feeding behaviour of the wood pigeon. *Behaviour* **40,** 10–42.

Nijhout, H. F., and Wheeler, D. E. (1982). Juvenile hormone and the physiological basis of insect polymorphisms. *Q. Rev. Biol.,* **57,** 109–133.

Petersen, B. (1954). Egg-laying and habitat selection in some *Pieris* species. *Entomol. Tidskr.* **75,** 194–203.

Phillips, P. A., and Barnes, M. M. (1975). Host race formation among sympatric apple, walnut, and plum populations of the codling moth, *Laspeyresia pomonella. Ann. Entomol. Soc. Am.* **68,** 1053–1060.

Phillips, W. M. (1977). Modification of feeding ''preference'' in the flea-beetle *Halitica lythri* (Coleoptera: Chrysomelidae) *Entomol. Exp. Appl.* **21,** 71–80.

Pietrewicz, A. T., and Kamil, A. C. (1979). Search image formation in the blue jay (*Cyanocitta cristata*). *Science* **204,** 1332–1333.

Pouzat, J. (1978). Host plant chemosensory influence on oogenesis in the bean weevil, *Acanthoscelides obtectus*. *Entomol. Exp. Appl.* **24,** 601–608.

Prokopy, R. J. (1972). Evidence for a marking pheromone deterring repeated oviposition in apple maggot flies. *Environ. Entomol.* **1,** 326–332.

Rankin, M. A. (1974). The hormonal control of flight in the milkweed bug *Oncopeltus fasciatus*. *In* "Experimental Analysis of Insect Behaviour" (L. Barton Browne, ed.), pp. 317–328. Springer–Verlag, New York.

Rankin, M. A. (1978). Hormonal control of insect migratory behavior. *In* "Evolution of Insect Migration and Diapause" (H. Dingle, ed.), pp. 5–32. Springer–Verlag, New York.

Rankin, M. A., and Rankin, S. (1980). Some factors affecting presumed migratory flight activity of the convergent lady beetle *Hippodama convergens* (Coccinellidae: Coleoptera). *Biol. Bull.* **158,** 356–369.

Rausher, M. D. (1978). Search image for leaf shape in a butterfly. *Science* **200,** 1071–1073.

Rausher, M. D. (1979a). Larval habitat suitability and oviposition preference in three related butterflies. *Ecology* **60,** 503–511.

Rausher, M. D. (1979b). Egg recognition: Its advantage to a butterfly. *Anim. Behav.* **27,** 1034–1040.

Rausher, M. D. (1979c). "Coevolution in a Simple Plant-Herbivore System." Ph.D. Dissertation, Cornell Univ., Ithaca, New York.

Rausher, M. D. (1980). Host abundance, juvenile survival, and oviposition preference in *Battus philenor*. *Evolution* **34,** 342–355.

Rausher, M. D. (1981a). The effect of native vegetation on the susceptibility of *Aristolochia reticulata* (Aristolochiaceae) to herbivore attack. *Ecology* **65,** 1187–1195.

Rausher, M. D. (1981b). Host selection by *Battus philenor:* the roles of predation, nutrition, and plant chemistry. *Ecol. Monogr.* **51,** 1–20.

Rausher, M. D. (1983). The ecology of host selection behavior in phytophagous insects. *In* "Variable Plants and Herbivores in Natural and Managed Systems" (R. F. Denno, and M. S. McClure, eds.), in press. Academic Press, New York.

Rausher, M. D., and Papaj, D. R. (1983a). Demographic consequences of conspecific discrimination by *Battus philenor* butterflies. *Ecology,* in press.

Rausher, M. D., and Papaj. D. R. (1983b). Foraging by *Battus philenor* butterflies: evidence for individual differences in searching behaviour. *Anim. Behav.,* in press.

Rawlins, J. E. (1980). Thermoregulation by the black swallowtail butterfly, *Papilio polyxenes* (Lepidoptera: Papilionidae). *Ecology* **61,** 345–357.

Ricklefs, R. E. (1973). "Ecology." Chiron, New York.

Robacker, D. C., and Ambrose, J. T. (1979). Effects of number of reinforcements and interference on visual and olfactory learning modalities of the honey bee (Hymenoptera: Apidae). *Ann. Entomol. Soc. Am.* **72,** 775–780.

Root, R. B., and Kareiva, P. (1983). The search for resources by cabbage butterflies (*Pieris rapae*): ecological consequences and adaptive significance of searching in a patchy environment. *Ecol. Monogr.,* in press.

Rothschild, M., and Schoonhoven, L. M. (1977). Assessment of egg load by *Pieris brassicae* (Lepidoptera: Pieridae). *Nature (London)* **266,** 352–355.

Saxena, K. N. (1969). Patterns of insect–plant relationships determining susceptibility or resistance of defferent plants to an insect. *Entomol. Exp. Appl.* **12,** 751–766.

Saxena, K. N. (1978). Role of certain environmental stimuli in determining the efficiency of host plant selection by an insect. *Entomol. Exp. Appl.* **24,** 666–678.

Schoonhoven, L. M., and Dethier, V. G. (1966). Sensory aspects of host-plant discrimination by lepidopterous larvae. *Arch. Neerl. Zool.* **16,** 497–530.

Shaw, M. J. P. (1970a). Effects of population density on alienicolae of *Aphis fabae* Scop. I. The effect of crowding on the production of alatae in the laboratory. *Ann. Appl. Biol.* **65,** 191–196.

Shaw, M. J. P. (1970b). Effects of population density on alienicolae of *Aphis fabae* Scop. II. The effects of crowding on the expression of the migratory urge among alatae in the laboratory. *Ann. Appl. Biol.* **65,** 197–203.

Shaw, M. J. P. (1970c). Effects of population density on alienicolae of *Aphis fabae* Scop. III. The effect of isolation on the development of form and behaviour of alatae in a laboratory clone. *Ann. Appl. Biol.* **65,** 205–212.

Singer, M. C. (1971). Evolution of food-plant preferences in the butterfly *Euphydryas editha*. *Evolution* **25,** 383–389.

Singer, M. C. (1982). Quantification of host preference by manipulation of oviposition behavior in the butterfly *Euphydryas editha*. *Oecologia* **52,** 230–235.

Singer, M. C. (1983). Determinants of muiltiple host use in a phytophagous insect population. *Evolution,* in press.

Slansky, F. (1980). Food consumption and reproduction as affected by tethered flight in female milkweed bugs (*Oncopeltus fasciatus*). *Entomol. Exp. Appl.* **28,** 277–286.

Stanton, M. L. (1979). The role of chemotactile stimuli in the oviposition preferences of *Colias* butterflies. *Oecologia* **39,** 79–91.

Stanton, M. L. (1982). Searching in a patchy environment: Foodplant selection by *Colias p. eriphyle* butterflies. *Ecology,* **63,** 839–853.

Stengel, M. (1974). Migratory behaviour of the female of the common cockchafer *Melolontha melolontha* and its neuroendocrine regulation. *In* "Experimental Analysis of Insect Behaviour" (L. Barton-Browne, ed.), pp. 297–303. Springer–Verlag, New York.

Swihart, C., and Swihart, S. (1970). Colour selection and learned feeding preferences in the butterfly, *Heliconius charitonius*. *Anim. Behav.* **18,** 60–64.

Tabashnik, B. E., Wheelock, H., Rainbolt, J. D., and Watt, W. B. (1981). Individual variation in oviposition preference in the butterfly, *Colias eurytheme*. *Oecologia* **50,** 225–230.

Thorpe, W. H. (1963). "Learning and Instinct in Animals." Methuen, London.

Tinbergen, L. (1960). The natural control of insects in pine woods. I. Factors influencing the intensity of predation in song birds. *Arch. Neerl. Zool.* **13,** 265–343.

Tinbergen, N., Impkoven, M., and Franck, D. (1967). An experiment on spacing out as a defence against predation. *Behaviour* **28,** 307–321.

Traynier, R. M. M. (1967). Stimulation of oviposition by the cabbage root fly *Erioischia brassicae*. *Entomol. Exp. Appl.* **10,** 401–412.

Traynier, R. M. M. (1979). Long-term changes in the oviposition behavior of the cabbage butterfly *Pieris rapae* induced by contact with plants. *Physiol. Entomol.* **4,** 87–96.

Truman, J. W. (1978). Hormonal control of insect behavior. *Horm. Behav.* **10,** 214–234.

Tsitsipis, J. A., and MIttler, T. E. (1976). Influence of temperature on the production of parthenogenetic and sexual females by *Aphis fabae* under short day conditions. *Entomol. Exp. Appl.* **19,** 179–188.

Vespalainen, K. (1971). Determination of wing length and alary dimorphism in *Gerris odontogaster*. *Hereditas* **69,** 308 (Abs.).

Vaspalainen, K. (1978). Wing dimorphism and diapause in *Gerris:* Determination and adaptive significance. *In* "Evolution of Insect Migration and Diapause" (H. Dingle, ed.), pp. 218–253. Springer–Verlag, New York.

von Frisch, K. (1967). "The Dance Language and Orientation of Bees." Belknap, Cambridge, Mass.

Waddington, K., Allen, T., and Heinrich, B. (1981). Floral preferences of bumblebees (*Bombus edwardsii*) in relation to intermittent versus continuous rewards. *Anim. Behav.* **29,** 779–784.

Waldbauer, G. P. (1978). Phenological adaptation and the polymodal emergence patterns of insects. *In* "Evolution of Insect Migration and Diapause" (H. Dingle, ed.), pp. 127–144. Springer–Verlag, New York.

Waldbauer, G. P., and Sternburg, J. G. (1973). Polymorphic termination of diapause by *Cecropia:* genetic and geographical aspects. Biol. Bull. **145,** 627–641.

Waloff, N. (1973). Dispersal by flight of leafhoppers (Auchenorrhyncha: Homoptera). *J. Appl. Ecol.* **10,** 705–730.

Wasserman, S. S., and Futuyma, D. J. (1981). Evolution of host plant utilization in laboratory populations of the southern cowpea weevil, *Callasobruchus maculatus* Fabricus (Coleoptera: Bruchidae). *Evolution* **35,** 605–617.

Wellington, W. G. (1957). Individual differences as a factor in population dynamics: the development of a problem. *Can. J. Zool.* **35,** 293–323.

Wellington, W. G. (1960). Qualitiative changes in natural populations during periods of abundance. *Can. J. Zool.* **38,** 289–314.

Wellington, W. G. (1964). Qualitative changes in populations in unstable environments. *Can. Entomol.* **96,** 436–451.

Wellington, W. G. (1965). Some maternal influences on progeny quality in the western tent caterpillar, *Malacosoma pluviale* (Dyar). *Can. Entomol.* **97,** 1–14.

White, R. R., and Singer, M. C. (1974). Geographical distribution of host plant choice in *Euphydryas editha* (Nymphalidae). *J. Lepid. Soc.* **28,** 103–107.

Wiklund, C. (1973). Host plant suitability and the mechanism of host plant selection in larvae of *Papilio machoan. Entomol. Exp. Appl.* **16,** 232–242.

Wiklund, C. (1974). Oviposition preference in *Papilio machoan* in relation to the host plants of the larvae. *Entomol. Exp. Appl.* **17,** 189–198.

Wiklund, C. (1975). The evolutionary relationship between adult oviposition preferences and larval host range in *Papilio machoan* L. *Oecologia* **18,** 185–197.

Yamamoto, R. T., Jenkins, R. Y., and McClusky, R. K. (1969). Factors determining the selection of plants oviposited by the tobacco hornworm, *Manduca sexta. Entomol. Exp. Appl.* **12,** 504–508.

# 4

# Spatial Patterns in the Plant Community and Their Effects upon Insect Search

MAUREEN L. STANTON



## I. INTRODUCTION

Herbivorous insects and their host plants are rarely distributed randomly on either a local or a geographic scale. A variety of processes can generate nonrandom dispersion patterns, including (1) limited mobility of seeds, pollen, and insects; (2) environmental differences affecting survival, growth, and reproduction; (3) competition among plants or among insects; and (4) interactions between plants and insects. In addition to these natural patterning processes, human agriculture creates highly concentrated, artificial patches of plants and of the insects that consume those plants. The ecological factors underlying nonrandom plant distributions are being extensively studied (Janzen, 1970; Harper, 1977), as are the roles played by spatial heterogeneity and competition in molding plant–herbivore communities (Connell, 1978). Furthermore, we know a great deal about specific interactions between individual plants and insect herbivores, as demonstrated by other chapters in this volume. We are only beginning to understand, however, how patterns of plant dispersion and diversity bring about



HERBIVOROUS INSECTS

ISBN 0-12-045580-3

the nonrandom distribution of herbivorous insects among their hosts. The assertion that plants are nonrandomly distributed implies that individual host plants grow in areas of varying floristic "texture," that is, that they are embedded within plant aggregations of varying density, dispersion, and species richness. This chapter, rather than asking how a "typical" host plant is located by insects, specifically examines how host plants growing within certain community types vary in their susceptibility to herbivore attack.

The mechanisms generating nonrandom herbivore distributions are not only important keys to understanding the evolution of plant–insect associations, but are also of great economic interest. A crop pest is an insect that reaches high enough densities upon economically important plants to reduce their commercial value. Whether pests arise in any given situation, however, depends on the texture of the host-plant community (Root, 1975), as well as on the behavior and physiology of the local insect fauna (Strong, 1979). For example, one often finds that plants growing in high-density, low-diversity aggregations (such as taiga forest and agricultural crops) are plagued by chronically high insect infestations (Pimentel, 1961a,b), whereas the same plant species suffer minimal damage when growing in complex, natural plant communities (Gibson and Jones, 1977). Patterns of this type can be understood only by identifying the processes whereby plant community texture influences the reproduction and behavior of insect herbivores. Some of these processes are examined by addressing four major questions:

1. Do changes in plant community texture have a consistent impact on the severity with which an individual host plant is attacked by herbivorous insects? If not, what can be learned from this apparent diversity of response?
2. What mechanisms can cause herbivores to become nonrandomly distributed among individual host plants or among host-plant areas of varying floristic texture?
3. Given a quantitative understanding of the cues an insect species uses to find host-plant localities (and to select individual plants), can one predict how the herbivore load on an individual plant will change as local plant community texture changes?
4. Can an understanding of insect searching mechanisms be used to design agricultural systems that are less likely to be discovered or exploited by potential pests?

Spatially heterogeneous herbivore infestation patterns can result from a wide variety of factors, many of which will be addressed in this chapter. Particular emphasis will be placed, however, on behavioral factors—that is, the interactions between plant community texture and herbivore searching behavior. Our knowledge of the interaction between plant spatial patterns and insect herbivore distribution comes almost exclusively from controlled experiments using agricultural-style plots (e.g., Pimentel, 1961a; Bach, 1980a; Kareiva, 1982). To counterbalance this bias, data from natural plant–insect communities are examined whenever possible.

## II. INSECT RESPONSES TO PLANT COMMUNITY TEXTURE

In natural communities, herbivores tend to be found in sites where the plants they use for nectaring, feeding, or egg laying are most abundant (e.g., Douwes, 1968; Brussard and Ehrlich, 1970; Singer, 1971; Copp and Davenport, 1978; Cullenward *et al.*, 1979). Conversely, plants that are widely dispersed or isolated from other members of their species often escape discovery by injurious insects (Janzen, 1971, 1972; Vandermeer, 1974; Davis, 1975). Studies like these imply that the herbivore load upon any given plant is somehow influenced by the texture of the surrounding community, but it is difficult to isolate the effects of density and diversity in natural plant assemblages (e.g., Simmons *et al.*, 1975). It is therefore necessary to discuss studies conducted in artificial communities of plants, where host-plant density, patch size, and species diversity can be manipulated independently.

On the basis of experimental studies of crucifers and their associated insect herbivores, Root (1973) outlined the "resource concentration" hypothesis. This hypothesis states that specialized herbivores are more likely to discover and exploit concentrated stands of host plants, resulting in increased damage to plants growing in dense monocultures. However, not all herbivore species respond in this way to host-plant patches (Tahvanainen and Root, 1972; Root, 1973; R. B. Root and P. Kareiva, unpublished data, 1982). A number of studies showing the effects of plant density and diversity on insect herbivores are summarized in Table I. In many cases, increased host-plant density results in heavier herbivore loads per plant, whereas increased plant diversity is associated with reduced herbivore attack. Still, the pattern is far from uniform. Several factors could contribute to this apparent diversity: (1) quantitative or qualitative differences in experimental design; (2) a lack of correspondence between density and diversity as perceived by the biologist and that which is important to the searching insect; and (3) variation in the host-plant orientation behavior of insects from different taxa. Each of these sources of variation is considered below.

Because the data in Table I were collected using a variety of experimental designs, it is often difficult to compare results among studies. Density manipulation studies have revealed all possible interactions (positive, negative, and neutral) between local host-plant abundance and herbivore load per plant. Insect populations may flourish in areas where food is plentiful and easy to locate.

**TABLE I.** Herbivore Responses to Host-Plant Density and Diversity[a]

| Scientific and Common Names[b] | Plant Diversity | Host-Plant Density | References |
|---|---|---|---|
| COLEOPTERA | | | |
| Chrysomelidae | | | |
| *Acalymma thiemei* (S) squash beetle | – | + | Risch (1980) |
| *Ceratoma ruficornis* (S) | – | + | Risch (1980) |
| *Diabrotica balteata* (G) banded cucumber beetle | = | + | Altieri *et al*. (1977); Risch (1981) |
| *Diabrotica adelpha* (G) | – | + | Risch (1980) |
| *Paranapicaba waterhousei* (G) | – | + | Risch (1980) |
| *Acalymma vittata* (O) striped cucumber beetle | – or + | = | Bach (1980b); Lower (1972) |
| *Phyllotreta cruciferae* (S) flea beetle | – | +**, =, –* | **Cromartie (1975a); *Pimetel (1961a,b); Tahvanainen and Root (1972); Root (1973) |
| *Phyllotreta punctulata* (O) flea beetle | – | = or –* | Root (1973); *Pimentel (1961a,b) |
| *Phyllotreta vittata* (O) flea beetle | – | = or –* | Root (1973); *Pimentel (1961a,b) |
| *Phyllotreta striolata* (O) striped flea beetle | – | + | Cromartie (1975a) |
| Coccinellidae | | | |
| *Epilachna varivestis* (O) Mexican bean beetle | – | + | Sloderbeck and Edwards (1979); Turner and Friend (1933) |
| Curculionidae | | | |
| *Anthonomus grandis* (S) boll weevil | n.d. | + | Parencia *et al*. (1964) |
| *Pissodes strobi* (O) white pine weevil | – | n.d. | Belyea (1923) |

| | | | |
|---|---|---|---|
| HOMOPTERA | | | |
| Cicadellidae | | | |
| *Empoasca kraemeri* (S) bean leafhopper | – | n.d. | Altieri *et al*. (1977) |
| *Empoasca fabae* (S) potato leafhopper | – | = | Mayse (1978); Kretzschmar (1948); Altieri *et al*. (1977) |
| Aphididae | | | |
| *Aphis craccivora* (O) | n.d. | – | A'Brook (1968) |
| *Aphis gossypii* (G) cotton aphid | n.d. | – | A'Brook (1968) |
| *Aphis fabae* (O) pea aphid | – | – | Smith (1976); A'Brook (1973); Way and Heathcote (1966) |
| *Brevicoryne brassicae* (O) cabbage aphid | – | = | Smith (1969); Pimentel (1961a); Root (1973); Theunissen and Den Ouden (1980); Dempster and Coaker (1974) |
| *Rhopalosiphum padi* | – | = or – | A'Brook (1973) |
| *Rhopalosiphum maidus* | | + | Mayse (1978) |
| *Myzus persicae* (G) green peach aphid | – | – | A'Brook (1973); Horn (1981) |
| Aleyrodidae | | | |
| *Aleyrodes brassicae* (S) white fly | – | n.d. | Smith (1976); Perrin and Phillips (1978) |
| ORTHOPTERA | | | |
| Acrididae | | | |
| *Melanoplus femurrubrum* (G) red-legged grasshopper | + | + | Sloderbeck and Edwards (1979) |

*(continued)*

**TABLE I.** *(Continued)*

| Scientific and Common Names[b] | Plant Diversity | Host-Plant Density | References |
|---|---|---|---|
| HEMIPTERA | | | |
| Lygaedae | | | |
| *Oncopeltus fasciatus* (S) large milkweed bug | n.d. | + | Ralph (1976); Sauer and Feir (1973) |
| Pentatomidae | | | |
| *Murgantia histrionica* (O) harlequin cabbage bug | n.d. | + | McLain (1981) |
| DIPTERA | | | |
| Chloropidae | | | |
| *Oscinella frit* (G) shoot fly | = | – | Adesiyun (1978); Mowat (1974) |
| Anthomyiidae | | | |
| *Erioischia brassicae* (O) cabbage root fly | – or = | – | Ryan *et al*. (1980); Dempster and Coaker (1974) |
| LEPIDOPTERA | | | |
| Pieridae | | | |
| *Pieris rapae* (O) cabbage butterfly | –*, =, +** | = or – | Root (1973); Theunissen and Den Ouden (1980); *Dempster (1969); **Cromartie (1975a); R. B. Root and P. Kareiva (unpublished data, 1982); Dempster and Coaker (1974) |
| *Colias alexandra* (S) | n.d. | – | Hayes (1981) |
| Heliconiidae | | | |
| *Agraulis vanillae* (S) gulf fritillary | n.d. | +? | Copp and Davenport (1978) |

| | | | |
|---|---|---|---|
| Papilionidae | | | |
| *Battus philenor* (S) pipevine swallowtail | – | n.d. | Rausher (1981) |
| Noctuidae | | | |
| *Mamestra brassicae* (O) cabbage moth | – | n.d. | Smith (1976); Theunissen and Den Ouden (1980) |
| Plutellidae | | | |
| *Plutella maculipennis* (O) diamondback moth | – | n.d. | Perrin and Phillips (1978); Smith (1976) |
| *Plutella xylostella* (O) diamondback moth | – | = | Buranday and Raros (1975); Theunissen and Den Ouden (1980); Cromartie (1975a) |
| Tortricidae | | | |
| *Choristoneura fumiferana* (O) spruce budworm | –? | n.d. | Simmons *et al.* (1975) |
| Oecophoridae | | | |
| *Depressaria pastinacella* (O) parsnip webworm | n.d. | – | Thompson and Price (1977) |
| HYMENOPTERA | | | |
| Cephidae | | | |
| *Cephus cinctus* (G) wheat stem sawfly | – | – | Smith (1976); Luginbill and McNeal (1958) |

[a]Qualitative patterns from a number of published studies are shown. A plus sign indicates a positive relationship between either density or diversity and the number of herbivores per host plant. Negative interactions are represented by a minus sign, and erratic or null relationships are shown by an equals sign. Note that where results differ between references, the relevant references are identified by asterisks.
[b](O), oligophagous species; (G), polyphagous species; (S), specialist (monophagous) species.

However, where host plants are extremely abundant, the finite population of herbivores in that area may become saturated in the short term, generating a negative relationship between host-plant density and herbivore load per plant. For example, Pimentel (1961a) found that *Phyllotreta* flea beetles were more numerous on sparsely planted *Brassica* host plants, but the dense plots in his experiment contained almost 400 times as many plants as the sparse plots. In other studies on chrysomelid beetles, where host-plant densities were varied only twofold, no differences in herbivore density were noted between treatments (Sloderbeck and Edwards, 1979; Bach, 1980b). One possible interpretation of these findings is that flea beetles respond in a fundamentally different way to host-plant density than the cucumber and bean beetles used in the latter two studies. It seems likely, however, that Pimentel's negative density effect resulted from short-term saturation of the local flea beetle populations. These examples illustrate the confounding of results that occurs when both herbivore type and experimental design are varied concurrently. Moreover, in short-term studies, quantitative differences in host-plant density may affect the qualitative relationship observed between plant abundance and herbivore load.

Some of the variation among the diversity manipulations summarized in Table I reflects the fact that plant diversity per se may be less important to a herbivore than its ability to exploit those apparently diverse species. In most cases, plants growing in monocultures are more heavily attacked than those growing intermingled with other plant species, but this phenomenon (associational resistance) appears to be most important for plants attacked by specialized herbivores (Root, 1973). For example, Risch (1980) showed that polyphagous chrysomelid beetles reached higher densities per plant in plots where corn, beans, and squash were mixed, whereas the more specialized beetles reached peak abundances where their host plant was growing alone. Similarly, cucumber beetles attacked cucumbers more heavily when these plants were planted next to a highly preferred squash cultivar (Lower, 1972). These studies imply that a plant's susceptibility to herbivore attack may actually increase if it is surrounded by acceptable host plants, even if they belong to different taxonomic groups. Thorsteinson (1960) noted that taxonomically diverse plants may share enough chemical or morphological traits to be perceived as a homogeneous group by herbivorous insects. To the extent that polyphagous herbivores orient to very generalized plant characteristics (such as foliage color or humidity), they may perceive diverse plant mixtures merely as dense host-plant stands. A specialized herbivore that responds to specific host-plant stimuli, however, may be repelled or confused by nearby nonhost plant species (e.g., Tahvanainen and Root, 1972). Thus, although mixed cropping systems can greatly reduce infestations of specialized herbivores (Altieri *et al.*, 1977), they may prove less effective against polyphagous pests (Sloderbeck and Edwards, 1979).

Because different insects orient to different cues while searching for host plants (Staedler, 1976), herbivore responses to plant community texture often vary along taxonomic lines. For example, most homopteran herbivores listed in Table I reach much higher densities per plant in sparse host-plant monocultures. Many species in this order are attracted to yellow-green light, and respond as strongly to long-wavelength light reflected from the earth's surface as to their host plants (Kennedy *et al.*, 1961; Saxena and Saxena, 1975). Densely planted hosts (or those surrounded by other vegetation) therefore attract relatively fewer leafhoppers or aphids than plants surrounded by bare soil (Way and Heathcote, 1966; Smith, 1969; Horn, 1981; but see Pimentel, 1961b). For somewhat different reasons, host plants silhouetted against bare ground may be more conspicuous to visually searching butterflies (Smith, 1976; Rausher, 1981), contributing to a negative relationship between plant density (or diversity), and the number of eggs laid upon each plant (R. B. Root and P. Kareiva, unpublished data, 1982). In contrast, herbivores that respond primarily to olfactory stimuli may find it easier to locate densely packed host plants. The large milkweed bug *Oncopeltus fasciatus* is attracted to host-plant volatiles (Ralph, 1976), and rarely colonizes small or sparse milkweed patches in the field (Ralph, 1977). Similarly, flea beetles that orient toward host-plant odors (Tahvanainen and Root, 1972) locate dense crucifer stands more frequently than sparse ones (Kareiva, 1982). In a later section of this chapter, some ways in which an insect's sensory capabilities influence its ability to exploit host-plant patches of varying size, density, and diversity will be explored.

## III. FACTORS GENERATING NONRANDOM HERBIVORE DISTRIBUTIONS

Until recently, the effect of plant spatial patterns on insect herbivores has been studied almost exclusively by periodic sampling of the insect fauna within experimental plots of varying size (Cromartie, 1975a), density (Pimentel, 1961a,b; A'Brook, 1968), and diversity (Root, 1973; Cromartie, 1975a). Sampling data of this type, however, do not identify the processes generating the observed nonrandom herbivore distributions (see Table II). Without studying the movements (Lawrence, in press) or survival (Bach, 1980a) of marked individuals, one does not know whether herbivore load is heavy in a given host-plant patch because insects were more likely to discover or remain within that area, or because local conditions for survival and reproduction are better there than elsewhere. I have divided the herbivore concentration mechanisms listed in Table II into two broad categories: (1) factors influencing the population growth of herbivores that successfully colonize a host-plant area; and (2) factors affecting the rate at which potential colonists immigrate into and emigrate from that area.

**TABLE II.** Factors Contributing to Localized Concentrations of Insect Herbivores

1. Increased success of residents
   a. Decreased starvation or dispersal risk
   b. Lower predation and parasitism
   c. Buffered microclimate
   d. Density-dependent host-plant quality
2. Increased number of colonists
   a. High herbivore immigration into area
      i. General attraction to habitat type
      ii. Attraction to widely occurring plant stimuli
      iii. Attraction to specific host-plant stimuli
   b. Limited herbivore emigration from area
      i. Recognition of patch boundaries
      ii. Area-restricted search after host-plant discovery
      iii. Physiological reduction in dispersal capability

### A. Plant Community Texture and Colonist Success

Herbivores will become locally abundant in areas where conditions favor better than average survival, growth, or reproduction. Once an insect arrives on a host plant, its survival probability is influenced not only by the plant upon which it feeds, but by the surrounding vegetation. Plant community texture can influence herbivore population growth in at least two ways: (1) by altering an insect's feeding rate or the benefit it derives from consuming host-plant material; and (2) by increasing or decreasing an insect's susceptibility to enemies.

#### *1. Quantity and Quality of Food*

Herbivore population growth in host-plant patches of varying density depends on the insect's dispersal capability and the host plant's developmental response to crowding. Especially for insects that have limited mobility, starvation risk and dispersal-related mortality may decrease in areas where host plants are closely spaced (Dethier, 1959; Gibson and Jones, 1977). For example, when young larvae of the pipevine swallowtail butterfly (*Battus philenor*) are forced to leave aging or defoliated host plants, few are able to locate plants more than a few meters away (Rausher, 1979). However, host-plant density may not be directly proportional to food density, because crowded plants often produce less leaf area and fewer reproductive parts than plants in less competitive situations (Harper, 1977). Although the parsnip webworm (*Depressaria pastinaca*) lays more eggs upon isolated (low-density) host plants, these plants produce many more of the closed umbels upon which larvae feed than do plants growing in dense clusters nearby (Thompson and Price, 1977).

Herbivore growth is dependent on the quality as well as the quantity of available host-plant material (e.g., Chew, 1975). Host-plant quality can in turn be influenced by plant community texture in a variety of ways. In natural communities, plants often become most abundant in areas where they grow best, potentially giving rise to a positive relationship between local host-plant density and the quality of food available to insects. However, in very dense agricultural fields, interplant competition can reduce the water and protein content of foliage (Milthorpe and Moorby, 1974), and could possibly result in decreased herbivore population growth (Scriber, 1977).

The quality of a herbivore's food may also be influenced by the proximity of nonhost plant species. Shading by taller plant species may lower foliage quality (Bach, 1981) or change the microclimate near the feeding herbivore (Price, 1976). Volatile chemicals from the leaves of surrounding nonhost species can reduce feeding rates in some specialist herbivores (Saxena and Prabha, 1977), and some plants may even acquire repellent compounds from neighboring nonhost species (Bach, 1981). There are also cases in which increased plant diversity actually enhances a herbivore's ability to consume its host plants. Risch (1980) found that some chrysomelid beetles preferred to eat diseased host-plant leaves, and that these leaves were more common in mixed plots than in monocultures. These varied textural effects on herbivore feeding and development can be teased apart, but the effort will require monitoring individual herbivores throughout their growth period under a variety of field conditions.

*2. Abundance of Enemies*

Lower numbers of insect herbivores in diverse plant stands have often been attributed to higher rates of predation or parasitism in weedy areas (e.g., Smith, 1969; Smith and Whittaker, 1980). Surrounding vegetation may alter a herbivore's susceptibility to enemies in at least two ways: (1) by providing alternative habitats for predators and parasitoids, and (2) by affecting the ease with which these animals locate their herbivore prey (Vinson, 1981). Pimentel (1961b) found that generalized predators such as ants and spiders were most abundant in diverse plots, whereas parasitoids that specialize upon crucifer feeders were most abundant in *Brassica* monocultures. Complex vegetation provides an excellent habitat for generalized insect predators, but parasitoids that orient specifically to the odor of their host's food plants (Read *et al.* 1970; Vinson, 1981) can be confused by volatiles from nonhost species (Monteith, 1960). Herbivores growing in mixed plant communities may therefore suffer from greater exposure to generalized predators though they are rendered less conspicuous to specialist parasitoids.

These patterns are suggestive, but studies of parasitism and predation rates in short-term agricultural plantings may be of limited relevance to natural plant–insect associations. One reason is that agronomic studies typically examine only the short-term responses of herbivore enemies to concentrations of their hosts. The rapidity with which enemies locate herbivore concentrations, however, may not reflect either their ability to multiply within such profitable areas or the equilibrium densities they can ultimately attain there. The insignificant levels of parasitism and predation mentioned in some plot studies (e.g., Tahvanainen and Root, 1972; Bach, 1980b) may simply reflect a time lag in the response of herbivore enemies to ephemeral host aggregations (Price, 1976). One sees much higher rates of herbivore predation or parasitism in plant communities that vary less dramatically over space and time (e.g., Hayes, 1981). In disturbed tropical habitats, Blau (1980) found that predators of the black swallowtail (*Papilio polyxenes*) were more numerous in older host-plant stands. A second reason is that some agronomic activities can be very disruptive to both herbivores and their enemies. It is easy to imagine that weeding (Mayse, 1978), tilling (Sloderbeck and Edwards, 1979), or mowing (Horn, 1981) could have greater impact on insect abundance than the variations in plant community texture these activities produce. A full understanding of the interplay between plant spatial patterns and herbivore susceptibility to enemies will require greater commitment to long-term manipulations in less disturbed environments.

## B. Plant Community Texture and the Number of Colonists

Host-plant patches that are especially conspicuous to searching insects will acquire a disproportionate share of the local herbivore population. A complete understanding of the spatial herbivore infestation patterns must therefore take into account the rate at which herbivores locate and colonize patches of different types (Price, 1976). Discussed in this section is the way in which plant community texture influences two aspects of insect searching behavior: (1) the rate at which host-plant patches are discovered; and (2) the intensity with which discovered patches are exploited. One can define *patch discovery* (or the immigration of a single potential colonist) as the moment an insect herbivore crosses some arbitrarily defined host-plant density isocline, traveling from lower to higher abundance. Emigration occurs when that density isocline is crossed in the opposite direction. Representing patches in this way allows one to apply heuristic models to insect searching movements, and, hence, to the relationship between herbivore load per plant and local plant spatial patterns.

### *1. Discovery of Host-Plant Patches*

Some insects are unable to orient toward rich host-plant aggregations using long-distance cues. Many species of aphids and thrips can be viewed as "aerial plankton"; they are carried by winds over enormous distances, and appear to be deposited at random with respect to host-plant distribution (Price, 1976). Other herbivorous insects can orient toward host plants under ideal conditions in the laboratory (see May and Ahmad, Chapter 6, this volume), but may be unable to respond to relevant stimuli in complex natural habitats (Robert and Blaisinger, 1978). The total number of plants encountered by a randomly moving herbivore depends only on the average density of host plants, and is unaffected by plant spatial distribution (for an analogous parasite–prey argument see Rogers, 1972).

Consider a herbivore that moves randomly through its habitat while searching for host-plant locations; it does not recognize host-plant patches per se. The probability that it immigrates into any host-plant area is directly related to the peripheral distance around that patch (or to the area of the patch, for insects that are deposited at random by high-altitude weather patterns). Larger patches will therefore be discovered more often than smaller patches, irrespective of host-plant density or species diversity within their boundaries. The expected herbivore load upon an individual plant can be determined only if we know how insects behave after patch discovery occurs. If, after locating a host plant, herbivores confine their movements to that immediate area, herbivory levels per plant will be strongly affected by local plant community texture. Herbivore searching behavior within host-plant areas will be dealt with in a following section. The aim here is to examine insects capable of orienting toward host-plant areas, and to ask how plant community texture could influence the rate at which they discover patches of different types.

Host plants growing in certain locations may be attacked disproportionately often because herbivorous insects are attracted to that area. To determine how plant community texture influences host-plant locality finding, it is necessary to address two questions: (1) at what distance does herbivore attraction occur? and (2) is the attraction specific to certain host plants, or is it a very generalized response to widely occurring plant stimuli? Although many insects respond to both ubiquitous and specific plant stimuli, specific orientation cues are most important to herbivores with a narrow feeding range (e.g., Thorsteinson, 1960; Prokopy and Owens, 1978). We should therefore expect to see consistent differences in the ways generalist and specialist herbivores respond to plant aggregations of varying texture (Root, 1973).

*a. Visual Search.* Many insect herbivores are attracted to plant shapes (Wallace, 1958; Copp and Davenport, 1978) or to vertical objects (Moericke *et al.*, 1975; Hamilton *et al.*, 1978). For tree-feeding insects, such as *Rhagoletis* flies, attraction to large vertical targets could conceivably operate at distances of many meters (Moericke *et al.*, 1975). Insects that feed upon open-field plants may fly toward treeless areas to find locations where their host plants are likely to occur (Cromartie, 1975b), and host plants growing in atypical plant communities may escape discovery (Chew, 1981). Long-distance orientation to plant shape is probably a very generalized response that can guide insects to likely habitats, but not to certain plant species. Once an insect has located a host-plant area, however, close-range shape recognition can become more specific, and more responsive to local plant community texture. Some butterflies recognize host plants on the basis of leaf shape (Papaj and Rausher, Chapter 3, this volume), and are less likely to find plants that are surrounded by dense vegetation (Rausher, 1981). Our knowledge of shape recognition in herbivores is too limited, however, to predict how local plant diversity could affect an insect's ability to find host-plant localities.

A searching insect may perceive an object's color at a greater distance than its shape. Shape recognition requires information from multiple ommatidia in an insect's eye, but color recognition can potentially occur when a single visual cell is stimulated (Wehner, 1975). Foliage-eating insects from many taxa are attracted to yellow-green hues (Kennedy *et al.*, 1961; David and Gardiner, 1962; Saxena and Saxena, 1975; Hamilton *et al.*, 1978; Saxena and Goyal, 1978), and the locust *Schistocerca migratoria* possesses visual receptors that respond specifically to peak leaf-reflectance spectra (Bennett *et al.*, 1967). These color-sensitive herbivores can probably detect dense vegetation at some distance and avoid fruitless exploration through sparsely vegetated areas (but see Kennedy and Fosbrooke, 1973). However, because the leaves of many plants reflect similar light wavelengths, attraction to greenish colors may not allow insects to discriminate among plant species (Kennedy *et al.*, 1961; Moericke, 1969). If visually searching herbivores cannot distinguish host plants from non hosts on the basis of leaf color, dense patches will attract potential colonists irrespective of the plant species they contain. Without knowing the extent to which insect visual orientation is species specific, or at what distances it can operate, we cannot predict how plant community texture will influence visually searching herbivores.

*b. Olfactory Search.* Visual stimuli are important to searching insect herbivores, but chemical cues play a far more conspicuous role in the evolution and maintenance of specific insect–plant associations (e.g., Thorsteinson, 1960; Ehrlich and Raven, 1964; Staedler, 1976). Once a plant is discovered, oviposition and feeding behaviors are tightly controlled by chemicals in host-plant leaves (see Feeny *et al.*, Chapter 2, and Hanson, Chapter 1, this volume). We are just beginning to understand, however, how chemical stimuli are involved in the search for host plants by insect herbivores. When air movement is kept to a minimum in the laboratory, insects can orient to plant odor gradients at distances of a few centimeters (Traynier, 1967; Schoonhoven, 1973; Saxena and Saxena, 1975), but under natural conditions, air turbulence quickly disrupts odor concentration gradients. Wherever there is directional airflow, an odor source can be located by a simple behavioral "rule": fly upwind so long as the attractive odor is discernible. This orientation mechanism (positive anemotaxis) allows many male insects to locate distant pheromone-emitting females (Schneider, 1974).

Targets baited with host-plant volatiles attract a wide variety of herbivores, including milkweed beetles (Ralph, 1976), apple maggot flies (Prokopy *et al.*,

1973), boll weevils (McKibben *et al.*, 1977), and bark beetles (Wood, 1973). These odor bait experiments do not reveal the mechanisms insects use to locate such targets, but positive anemotaxis seems likely. The desert locust *Schistocerca gregaria* moves rapidly upwind in an air current passed over grass leaves (Kennedy and Moorehouse, 1969), but drifts downwind when exposed to "clean" air (Haskell *et al.*, 1962). Female cabbage root flies (*Delia brassicae*) fly upwind and land frequently until they pass a source of crucifer odor, then turn downwind and reenter the plume (Hawkes *et al.*, 1978). Female *Agraulis* butterflies (Copp and Davenport, 1978) and Colorado potato beetles (May and Ahmad, Chapter 6, this volume) also orient upwind to the odors of their preferred host plants. Laboratory experiments like these must be interpreted cautiously, but the elegant experiments of Hawkes (1974) unequivocally demonstrate that insects can use positive anemotaxis to locate host-plant areas under field conditions.

The geometry of odor plumes helps one understand how plant spatial patterns might influence the behavior of insects orienting upwind to host-plant odor. Sutton (1949) developed a general model to calculate the average concentration of airborne molecules downwind from a point source. The model includes complex terms describing gustiness and ground roughness, and is extremely sensitive to changes in wind speed. At low wind speeds, odor concentration decreases as a complex exponential with increasing downwind distance. For "average" conditions (Wright, 1958) and a windspeed of 81 cm/sec (2 mph), the expected odor concentration ($C$) at a point $d$ cm directly downwind from a point source is estimated as follows:

$$C = S \times 2 \times 10^3 \times d^{-1.75} \times e^{-d^{-1.75}} \qquad (1)$$

where $S$ is the odor concentration at the source. For purposes of this analysis, assume that an insect moves upwind when it perceives host-plant odors above some threshold concentration $T$, but moves randomly at other times. By calculating how far downwind above-threshold odor concentrations extend, it is possible to estimate the "attractive area" for a host-plant patch—that is, the downwind area within which herbivores will orient to the patch. It is assumed that the number of times a plant is discovered is roughly proportional to the attractive area of the patch divided by the number of plants inside the patch. Odor concentrations downwind from three hypothetical host-plant patches were calculated using Eq. (1), and by making the simplifying assumption that convection due to wind has a much greater impact on odor molecule movement than diffusion (Fig. 1).

The olfactory attractiveness of a patch of host plants changes with patch size and density, and varies substantially for insects possessing different response thresholds. Patches (a) and (b) of Fig. 1 are the same size, but the second contains twice as many plants as the first. The concentration of host-plant volatiles may be greater

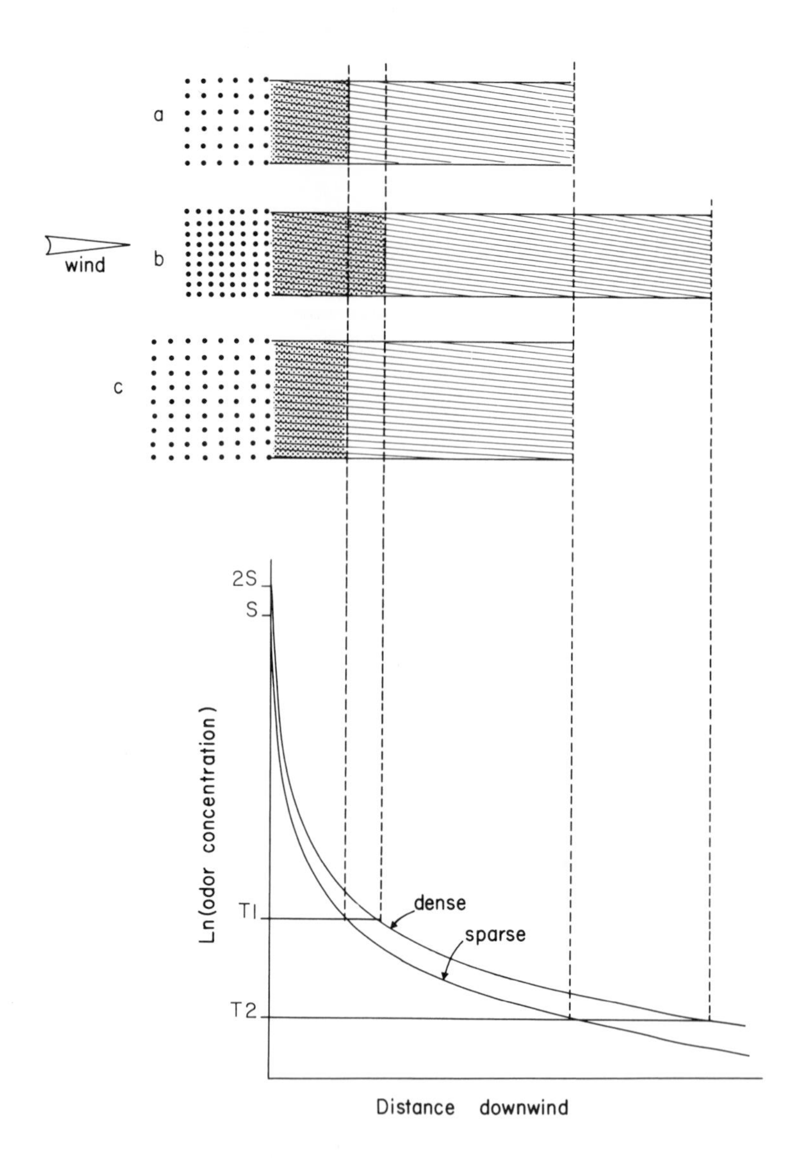


**Fig. 1.** Olfactory attractiveness of host-plant patches. Odor concentration downwind from patches of two densities are graphed at the bottom of the figure. The low-density odor curve applies to Patches (a) and (c), whereas the high-density curve applies to Patch (b). The curves represent an idealized situation in which diffusion is overshadowed by convection due to wind. In still air, odor concentration cannot be changed by altering host-plant density. Actual odor concentration depends on the relative importance of convection and diffusion, but can never be twice as great over the dense patch than over the sparse patch. Incorporating diffusion into this model magnifies the effect described in the text—i.e., detection area for each host plant will decrease as stand density increases. Downwind ''attractive areas'' are shown as rectangles although they are actually ellipsoid in shape (Wright, 1958). The attractive areas for the low-sensitivity herbivore (threshold T1) are stippled; those for the high-sensitivity herbivore (T2) are shaded.

at any given distance downwind from the dense patch, but because of both diffusion and the slope of the odor convection curve, the attractive area for Patch (b) is never twice that of Patch (a) (also see Bossert and Wilson, 1963). Even discounting diffusion, Patch (b) only attracts about half again as many herbivores as Patch (a), hence the attractive area per plant will be less than that in the sparse patch. If this is a viable interpretation of long-distance olfactory search by insects, plants growing in dense stands may attract fewer herbivores. Patch (c) contains the same number of plants as Patch (b), but is twice as large. Odor concentration downwind from Patch (c) will be less than that downwind from Patch (b), but the capture area is broader because the larger patch intersects the wind over a greater distance. The combined effects of diffusion and convection increase the attractive area of each plant in Patch (c) relative to Patch (b).

The olfactory sensitivity of searching herbivores also influences the expected attack rate upon plants growing in different patches. When a host plant is sought out by an insect that can respond to very low odor concentrations (T2 in Fig. 1), its attractive area can increase dramatically with a doubling of sensitivity, although a dense patch will still attract fewer herbivores per plant than a sparse patch of the same size. For insects that can detect only highly concentrated host-plant odors (near T1 in Fig. 1), a doubling of receptor sensitivity will not increase attractive area as much as for insects with greater olfactory sensitivity. Conversely, changes in patch structure should have greatest impact on those insects with keenest odor sensitivity. To predict the effect of host-plant distribution on patch-discovery rate, one must take into account the quantitative relationship between plant density, patch size, and insect olfactory sensitivity.

I know of no studies that have attempted to quantify behavioral response thresholds in a herbivorous insect to host-plant odor. Some male moths can detect female pheromone odor at concentrations of $10^{-16}$ g/cm$^3$ (Schneider, 1974), allowing them to orient upwind to females at distances of 2 miles or more (Bossert and Wilson, 1963), but it seems doubtful that herbivorous insects could be this sensitive to host-plant volatiles (Hanson, Chapter 1, this volume). Although boll weevils eventually discover traps baited with cotton oil up to 4 km away from the nearest cotton field (McKibben *et al.*, 1977), upwind orientation to host-plant odor has not been proven for distances greater than 15 m (Hawkes, 1974). It would be very interesting to know whether variation in the colonization ability of different insect herbivores (e.g., Davis, 1975) reflects varying sensitivity to host-plant volatiles. Such information could provide an essential link between the fields of sensory physiology and behavioral ecology.

Herbivore abundance is generally less in mixed plant stands than in monocultures (Table I), and it is reasonable to ask whether a plant surrounded by nonhost species is less conspicuous to odor-sensitive insect herbivores than one growing in a pure stand (Root, 1973). In field experiments using pure and mixed plant gardens, it is often difficult to tell whether nonhost plants interfere with olfactory search, or simply function as physical barriers to insect herbivores (e.g., Dempster, 1969; Ryan *et al.*, 1980). In some cases, however, it is clear that olfactory stimuli from nonhost plants confuse or repel specialist insects. Tahvanainen and Root (1972) found that *Phyllotreta* flea beetles reached lower densities upon kale plants growing near meadow vegetation (or planted with tomato seedlings) than upon those in pure stands of comparable density. Their laboratory tests showed that tomato plant odor interfered with the ability of flea beetles to locate their crucifer hosts. Similarly, Altieri *et al.* (1977) found that densities of the tropical bean beetle *Empoasca kraemeri* could be greatly reduced by leaving a grassy border around bean fields. In the laboratory, the ability of *E. kraemeri* to locate hidden bean foliage was inhibited by placing grass cuttings with the host-plant material. Potato beetles (*Leptinotarsa decemlineata*) orient upwind to the odor of solanaceous plants, but some nonhost species actually repel them (Visser and Nielson, 1977), and it would not be surprising to see lower densities of *L. decemlineata* in plots where their hosts are interplanted with repellent species.

Increasing local plant diversity should have its most pronounced effect on specialized herbivores that orient to specific host-plant odors. Insects that respond to ubiquitous plant volatiles and water vapor (Saxena and Saxena, 1975) may be attracted to densely vegetated areas regardless of the species they contain (e.g., Risch, 1980). Similarly, mixing closely related plants together may increase an area's attractiveness to oligophagous insects if the component plant species all release similar attractive volatiles (Lower, 1972). This subtle interplay between plant chemical diversity and insect olfactory selectivity may help explain why intercropping has such varied effects on herbivores within a single location (e.g., Theunissen and Den Ouden, 1980). To understand the effects of plant community texture on immigration processes in herbivorous insects, both in natural and in agricultural systems, we need considerably more data on the ability of insects to code complex visual and chemical information (Huber, 1978; Dethier, 1980).

### 2. *Searching Behavior after Patch Discovery*

Herbivore densities inside any host-plant area depend not only on how many insects discover the patch, but also on the average number of plants attacked by each immigrant. In the last section it was argued that the olfactory attractiveness of a host-plant patch does not increase as fast as its density, hence a dense stand should attract fewer insects per plant than a sparse patch of the same size and shape. In contrast, there are numerous reported cases showing that dense host-plant areas acquire a disproportionate share of herbivore colonists. This apparent discrepancy may arise because herbivores do not search randomly once a host-plant site is located, but instead restrict their movements to profitable areas. Searching behavior within host-plant patches can be divided into two basic types: (1) uniform searching, where the rules governing an insect's movements are unaffected by host-plant encounters; and (2) patch-restricted searching, where host-plant discovery leads to systematic reduction in dispersal. Insects using these two techniques will respond very differently to variations in plant community texture.

*a. Uniform Searching.* Two extreme examples of uniform searching within host-plant patches of varying density are shown in Fig. 2. In the first case (Fig. 2a), the herbivore flies some constant or randomly varying distance after visiting one host plant before searching for another. The insect encounters about four times as many plants in the dense patch as in the sparse patch, but flies over many of these without landing. Because it traverses the same distance in the two patches, it lands the same number of times in each and attacks proportionately fewer plants in the dense area. In the second case (Fig. 2b), the herbivore lands upon each host plant it encounters with constant or randomly varying probability. So long as time is not limiting, it encounters and attacks about four times as many plants in the dense patch, and the probability that an individual plant is attacked is independent of patch density.

Some herbivores respond to host-plant density in a way that is intermediate between these two models. The cabbage butterfly (*Pieris rapae*) often lays more eggs per host plant at low crucifer densities (Cromartie, 1975a; R. B. Root and P. Kareiva, unpublished data, 1982). By watching the movements of individual females, Root and Kareiva showed that this negative relationship between host-plant density and eggs per plant resulted from three aspects of the females' searching behavior: (1) patches of varying density were discovered at the same rate; (2) females landed only slightly more often in dense host-plant patches; and (3) egg-laying flights were highly directional, causing females to emigrate quick-

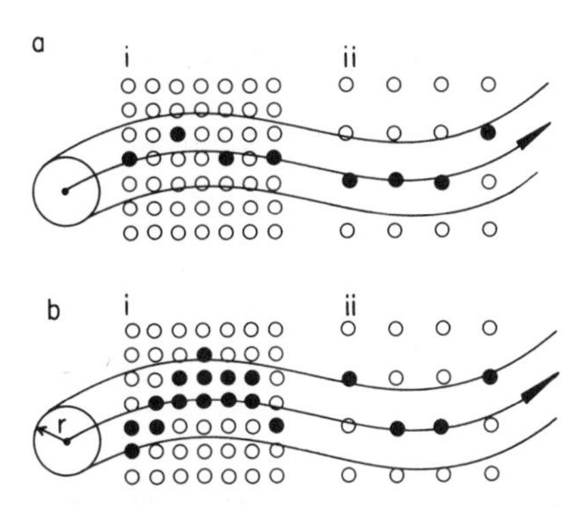


**Fig. 2.** Uniform search within host-plant patches. In each case, the herbivore flies a wandering path from left to right, and perceives all host plants within its perceptual radius $r$. In (a), the herbivore lands upon the first host plant it encounters after flying some constant distance $d$ from its previous landing. In (b), the insect lands upon each perceived host plant with $p = 0.5$. The plants selected are shown by shaded circles. Patch i is approximately four times denser than Patch ii.

ly from the patches they discovered. Similarly, egg-laying *Colias* butterflies land more often in areas where their preferred oviposition plants are abundant, but their landing rate reaches an upper limit in areas where host plants are very dense (Fig. 3). In the densest host-plant patches, *Colias* searching movements are essentially independent of local host-plant abundance (as shown in Fig. 2a). The searching movements of both *Pieris* and *Colias* butterflies effectively spread out

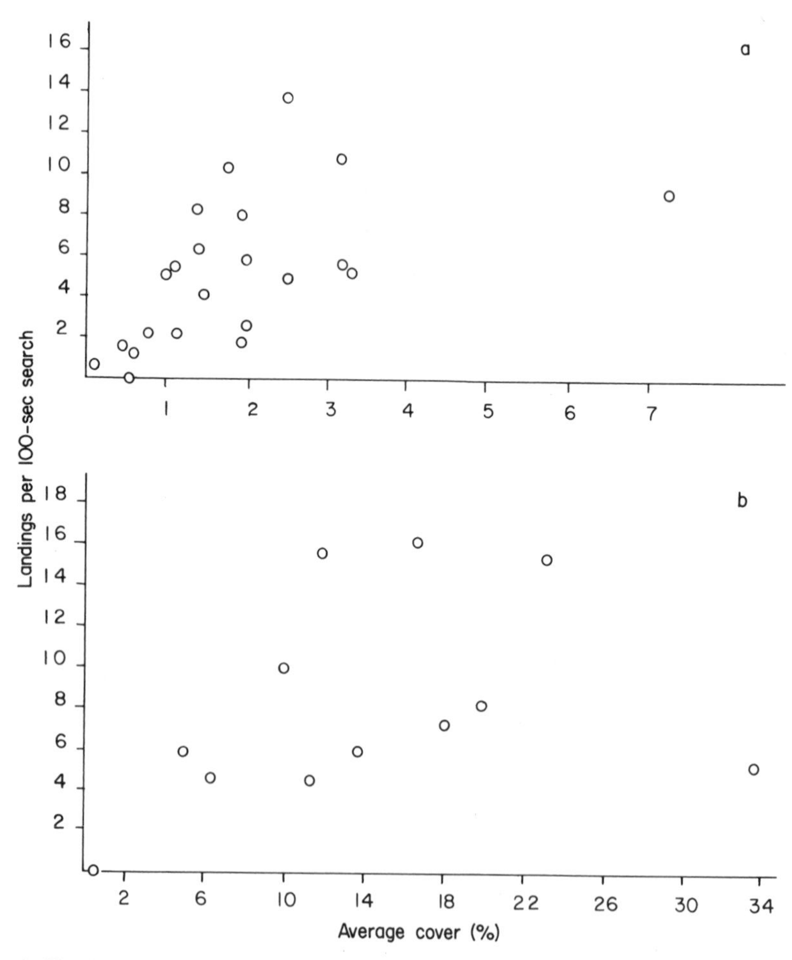


**Fig. 3.** The landing rate of egg-laying *Colias* butterflies as a function of preferred host-plant abundance. Each point represents a behavioral sequence of at least five host-plant landings. Search time does not include time spent on plants or inactive periods. (a) *Vicia americana*, $r = 0.5087$, $N = 22$; (b) *Trifolium hybridum*, $r = 0.2686$, $N = 12$. Percentage cover of the two preferred legume host plants was estimated from quadrats sampled along each female's flight path.

their reproductive effort over space, a tactic that may have evolved in response to spatially unpredictable larval survivorship (R. B. Root and P. Kareiva, unpublished data, 1982). Herbivores that are less mobile than these butterflies may benefit more by fully utilizing a host-plant patch when it is encountered.

Uniform searching behavior can result in either a negative or a null correlation between host-plant density and herbivore load per plant, but cannot generate greater herbivore numbers per plant in dense patches unless other factors are involved. Some insects may encounter plants at random, but select only those growing under conditions favorable for growth and survival. The seed-eating milkweed bug *Oncopeltus fasciatus* prefers to lay its eggs where host-plant pods are locally abundant (Sauer and Feir, 1973; Ralph, 1977). Harlequin cabbage bugs (*Murgantia histrionica*) and shoot flies (*Oscinella frit*) also select dense host-plant stands for oviposition, where the survival of their offspring is greatest (Adesiyun, 1978; McLain, 1981). In contrast, some herbivores prefer to lay eggs upon isolated plants, possibly because of decreased parasitism risk there (Read *et al.*, 1970). Nonrandom distribution of herbivores (or their eggs) thus does not imply that searching movements are nonrandom, because choices made after host-plant discovery may be influenced by local plant spatial patterns.

*b. Patch-Restricted Searching.* Where host plants are rare or unpredictable in space, a herbivore's reproductive success may depend on its ability to locate and thoroughly utilize scattered host-plant patches (Southwood, 1962; den Boer, 1970; Blau, 1980; Denno *et al.*, 1980). A number of herbivores have evolved elaborate (and rather inflexible) means to remain in rich host-plant locations. For example, when the seed bug *Neacoryphus bicroris* arrives in a dense host-plant patch, its flight muscles degenerate to leave further space for ovaries (Solbreck and Pehrson, 1979). In many aphid species, the development of alate dispersal forms is inhibited so long as food is plentiful (Kennedy and Fosbrooke, 1973). Many insects use behavioral mechanisms to remain within "profitable" sites, and one can now ask how plant community texture might influence the rate at which such herbivores emigrate from host-plant patches.

Some studies have shown that herbivorous insects move freely within a host-plant patch, but rarely migrate between patches (Tabashnik, 1980; Kareiva, 1982; Lawrence, 1983). One explanation for this pattern is that searching insects recognize the edge of a patch and turn back on encountering it. By marking *Tetraopes* milkweed beetles within a naturally occurring host-plant patch, Lawrence (1983) showed that individuals moved randomly in the interior of the stand, but tended to change direction at its edge. Although beetles frequently traversed the length of the large milkweed patch, they rarely moved into the surrounding meadow vegetation. Such area-restricted search may enhance both the egg-laying efficiency of females and the mating success of males.

Edge recognition may be feasible only in situations where host-plant patches possess conspicuous boundaries. For example, Kareiva (1982) found that *Phy-*

*llotreta* flea beetles readily moved between crucifer clumps along the length of a single plowed strip, but rarely crossed the diverse meadow vegetation that separated adjacent strips by only a few meters. Ralph (1976) observed that milkweed bugs (*Oncopeltus fasciatus*) remained in milkweed patches with sharp boundaries, but tended to emigrate from stands where milkweeds gradually blended into the surrounding vegetation. These observations suggest that patch edge recognition may be a more feasible dispersal-reducing mechanism in agricultural situations than in natural communities, where many plants, although clumped in their distribution, do not form dense, monospecific stands (Kershaw, 1975).

The impact of plant spatial patterns on patch edge recognition can be examined by comparing insect movements within agricultural fields to those observed in natural plant communities. The sulfur butterfly *Colias philodice eriphyle* is found in two distinct habitats in southern Colorado. It is a persistent pest of alfalfa crops at low elevations, but feeds upon legumes in mixed grassland communities at high elevations. Mark–recapture studies have shown that male and female *C. philodice eriphyle* typically disperse less than 100 m in alfalfa fields (Tabashnik, 1980), but travel twice that distance between recaptures in a mixed grassland community 150 km away (Watt *et al.*, 1979). Preliminary data suggest, however, that egg-laying females make much longer single flights in the alfalfa fields than in the grassland site (Stanton, in press; B. E. Tabashnik, personal communication). Females move rapidly within alfalfa fields, but rarely emigrate beyond field boundaries because they turn abruptly at the conspicuous crop borders (B. E. Tabashnik, personal communication). In the grassland population, legume host plants form much less obvious patches, as shown by the transect sample illustrated in Fig. 4. Legume species are significantly clumped in this area (Stanton, 1982), but the patches are typically quite sparse, and contain large numbers of nonlegume species. Searching females do not appear to recognize patch boundaries in the grassland, and their overall dispersal is less tightly restricted to host-plant patches than that of females in pest populations.

Where host plants do not grow within discrete, easily recognized stands, insects may use techniques other than edge recognition to remain within profitable areas. After encountering food, prey, or hosts, many insects exhibit intensive search behaviors: some combination of increased turning rate, decreased movement length, or increased searching "thoroughness" (Bond, 1980 and references therein). These behaviors tend to keep foraging animals in areas where their resources are locally dense, allowing sparser areas to escape exploitation (Murdie and Hassell, 1973). Because area-restricted searching can increase an animal's foraging efficiency (Rogers, 1972) and stabilize predator–prey interactions (Rogers and Hubbard, 1974; Oaten, 1977), it has received considerable theoretical attention. Most of this theory has focused on parasites and predators, but Kareiva (1983) has recently explored the effects of host-plant distribution on searching efficiency and population growth in insect herbivores.

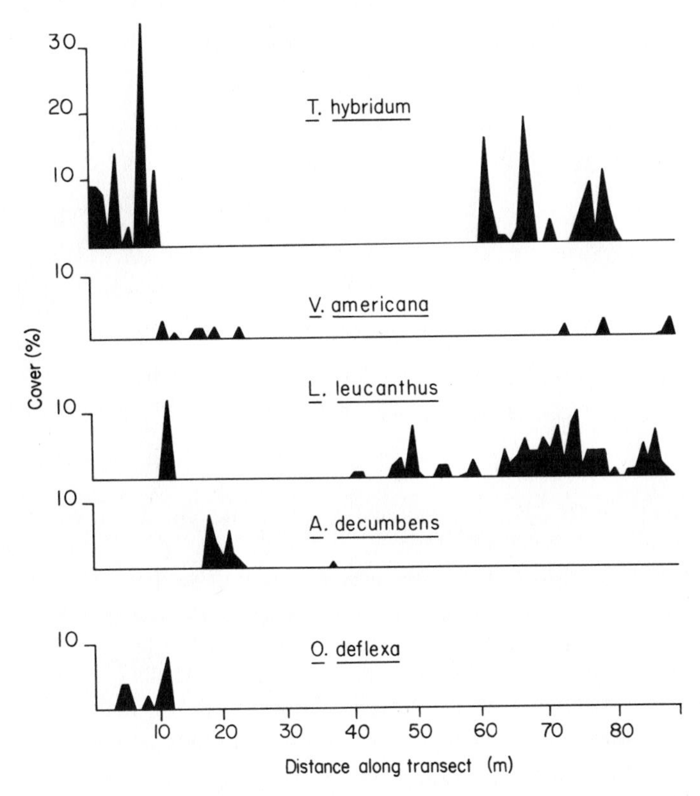


**Fig. 4.** Distribution of *Colias* host plants in a natural community. Quadrats were placed at 1-m intervals along six transects through a submontane grassland, of which one transect is illustrated here. The estimated percentage cover of five legume host plants is graphed on separate horizontal axes.

To the extent that host plants are patchily distributed and that thorough patch utilization enhances fitness, we should expect to see intensive searching behavior in herbivorous species. A large number of parasitic (Vinson, 1981) and predatory (Bond, 1980) insects respond to prey encounters by increasing the rate at which they change direction, thereby tending to remain in that area. Comparable data on herbivore searching movements are rare. Golik and Pienkowski (1969) showed that the odor of alfalfa leaves more than doubled turning rates in adult alfalfa weevils (*Hypera postica*). Egg-laying *Cidaria albulata* moths tended to turn more often inside experimental plots baited with the odor of their *Rhinanthus* host plants (Douwes, 1968). Jones (1977) showed that female *Pieris* butterflies carrying large numbers of mature eggs increased their turning rate within plots where host plants were abundant. These observations suggest at least

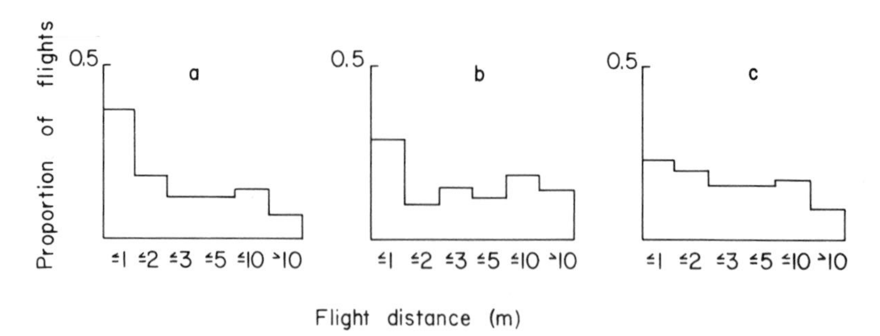


**Fig. 5.** Distances flown by *Colias* butterflies after visiting different plant types. (a) Flights after landings on preferred legume host plants ($N$ = 337). (b) Flights after landings on less preferred host plants ($N$ = 115). (c) Flights after landings on nonhost-plant species ($N$ = 216).

two ways in which plant community texture could affect the rate at which herbivores emigrate from host-plant areas. First, if turning rate is an increasing function of local host-plant density (because of concentrated odors or frequent host encounters), a herbivore may use a greater proportion of plants growing in dense stands than in sparse areas. This phenomenon has been documented in some parasites (Rogers, 1972), but comparable data regarding searching movements in insect herbivores are not yet available. Second, if encountering nonhost plant stimuli increases movement length or reduces turning rate, herbivores will tend to emigrate from diverse plant areas sooner than from host-plant monocultures.

Local plant diversity can have a pronounced effect on herbivore emigration rate. Bach (1980a,b) found that striped cucumber beetles (*Acalymma vittata*) reached much higher densities in cucumber monocultures than in adjacent mixed plots. She observed no significant differences in immigration, predation, or parasitism between diversity treatments, but found that beetles were less likely to leave pure host-plant stands. She suggested that beetles flew further after landing upon nonhost plants, causing emigration from diverse areas (also see Risch, 1981). The flight behavior of *Colias* butterflies lends support to this hypothesis (see Fig. 5). Females strongly prefer certain legume species for oviposition, and fly shorter distances after landing on these species than after visiting either less preferred legumes or totally unacceptable plants ($G = 22.730$; 10 df; $p < 0.05$). Because females were more likely to land upon nonhost plants where legumes were locally unavailable (M. L. Stanton, unpublished data), their flight behavior caused them to linger within patches of preferred hosts, but move rapidly through unprofitable areas. If other insects have evolved similar searching mechanisms, mixing host plants with nonhost species may decrease the amount of time that potential pests spend within agricultural fields.

## IV. IMPLICATIONS FOR AGRICULTURAL SYSTEMS

Understanding the processes underlying nonrandom herbivore distributions is crucial if we are to design agricultural systems that are less vulnerable to herbivore outbreaks than the expansive monocultures now used in most temperate regions (Risch, 1980). Heavy herbivore accumulations occur for a variety of reasons: (1) favorable local conditions for survival, growth, and reproduction; (2) increased herbivore immigration into that site because of attractiveness to searching insects; and (3) low emigration or dispersal from the area. The degree to which an individual host plant is attacked depends on the quantitative balance among these factors in any patch. Unfortunately, our understanding of how plant spatial patterns influence herbivore dynamics is both spotty and qualitative.

Most of the data on this problem come from studies in which insects were sampled periodically within artificial patches of plants of varying size, density, and diversity. Such studies provide groundwork for further research on spatially heterogeneous herbivore densities, but they often confound the effects of herbivore mortality, immigration, and emigration. For example, Lewis and Waloff (1964) found that mirid bugs disappeared quickly from the edge of a host-plant stand, but were fairly sedentary in the center of the plot. They ascribed this difference to emigration from the patch edge, but it may also reflect higher predation rates at the plot boundaries (e.g., van Emden, 1965). Similarly, Smith and Whittaker (1980) observed that *Gastrophysa* beetles disappeared more rapidly from host plants surrounded by mixed vegetation than from those in monocultures. Their results may be due either to increased predation in the diverse plot (as they suggest) or to increased emigration from that area (e.g., Bach, 1980b). As a few hypothetical scenarios can illustrate, it is important to determine how plant community texture influences insect host-plant search. The examples provided next make no pretense to realism; they are intended to be heuristic rather than simplistic.

Simple changes in the orientation or shape of a host-plant patch could affect that patch's susceptibility to herbivore attack. Consider the two rectangular patches shown in Fig. 6a. They are the same shape, contain the same number of host plants, and would accumulate the same number of randomly searching herbivores. However, for insects that orient upwind to host-plant odor, Patch (a2) has a smaller attractive area because its narrow dimension is perpendicular to the prevailing wind. This patch would attract fewer odor-sensitive herbivores than Patch (a1). Now consider how a change in patch shape could alter the balance between herbivore emigration and immigration. The two patches in Fig. 6b contain equal numbers of host plants, but the elongated patch (b1) has a greater perimeter : area ratio. Randomly searching herbivores will enter Patch (b1) more often than Patch (b2). If insects continue to move randomly inside a patch, they would tend to leave Patch (b1) sooner than (b2), and per-plant

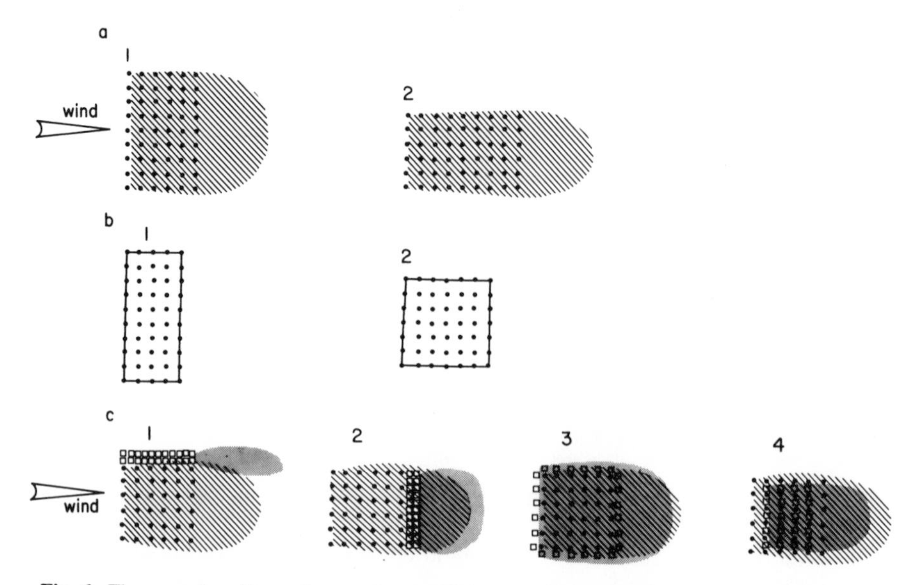


**Fig. 6.** The geometry of host-plant patch susceptibility. (a) The effect of patch orientation relative to prevailing wind. (b) The effect of patch shape. (c) Four mixed cropping arrangements relative to prevailing wind. Host plants are represented by circles, and their odor plumes are shaded; nonhost plants are represented by squares with stippled odor plumes.

herbivore loads would be equivalent in the two patches. But if these immigrants change direction at patch borders, the elongated patch would retain them about as long as the square patch, leading to potentially higher herbivore densities in Patch (b1).

Similar arguments suggest that the attractiveness of different intercropping arrangements could vary according to an insect's host-plant orientation mechanisms. In Fig. 6c, four possible arrangements of host plants (circles) and nonhost plants (squares) are shown. It is assumed that the nonhost species reduces herbivore attack in two ways: by chemically masking attractive host-plant odors and by physically blocking insect access to hosts. Arrangement (c1) would have little deterrent effect on insects that orient upwind to host-plant odor, because the odor plume of the nonhost plants barely overlaps with that of the host plants. In contrast, placing nonhosts slightly downwind from the host-plant patch—as in arrangement (c2)—could inhibit upwind orientation by insect herbivores. The solid nonhost border shown in arrangement (c3) would reduce immigration by odor-sensitive herbivores, and also might provide a habitat for generalist predators. However, insects that did manage to locate this host-plant patch might remain there longer if the bordering plants inhibit emigration as well as immigration. In contrast, the mixed planting arrangement (c4) might confuse olfactory searchers without reducing (and possibly increasing) emigration from the patch. This would be an effective planting strategy, if not negated by interspecific competition (e.g., van Emden and Williams, 1974).

At the present time we do not possess enough information about herbivore searching behavior to evaluate scenarios like those outlined. A full understanding of spatial nonrandomness in herbivore populations requires that more attention be focused on the behavior and success of individuals. Revealing immigration and emigration processes will require measuring the movements of individual herbivores within host-plant patches of varying size, density, and diversity. Once we understand the ways in which colonists discover and remain in different patch types, we can ask how the success of those immigrants is influenced by local plant community texture. Such detailed data on individual herbivores within both agricultural and natural plant communities will greatly enhance our understanding of plant–herbivore dynamics and pest outbreaks.

## V. SUMMARY

A number of studies have shown that host plants growing in dense, monospecific stands accumulate more herbivores than those growing in mixed vegetation. This pattern is most pronounced for specialist herbivores, suggesting that the specificity of an insect's host-plant-finding mechanisms influence its response to plant community texture. Herbivores may be attracted to dense vegetation by visual cues (foliage colors or plant shapes), but long-range visual orientation is probably less sensitive to plant community texture than olfactory search. To the extent that nonhost plant odors can repel specialist herbivores or mask attractive odors emanating from host plants, diverse patches will attract fewer immigrants than pure host-plant stands. By assuming that insect herbivores use positive anemotaxis to locate distant host plants, it is shown that dense host-plant patches will attract relatively fewer herbivores per host than sparse patches of the same size. Higher herbivore numbers per plant in dense patches are probably due to factors other than immigration—for example, increased feeding efficiency, decreased predation, or decreased emigration. Like many parasites and predators, insect herbivores probably possess behavioral mechanisms that increase their ability to remain within discovered host-plant areas. The effectiveness of such mechanisms (including patch edge recognition and host-induced turning) will depend on plant community structure. In particular, encounters with nonhost plants may increase the rate at which herbivores emigrate from host-plant areas. More data on insect searching movements relative to plant community structure are needed to uncover the processes underlying nonrandom herbivore dispersion patterns.

## Acknowledgments

I wish to thank S. Handel, R. Grosberg, and W. Lawrence for their thoughtful criticisms of the manuscript. P. Kareiva, R. Root, B. Tabashnik, and W. Lawrence graciously shared ideas and unpublished work. Conversations with L. Buss, L. Gall, T. Seeley, and S. Handel helped formulate many of the thoughts presented here, and H. Hasbrouck provided emergency assistance when it was needed most. This chapter has been greatly improved by the help of all those mentioned, but its flaws are entirely the author's responsibility.

## REFERENCES

A'Brook, J. (1968). The effect of plant spacing on the numbers of aphids trapped over groundnut crop. *Ann. Appl. Biol.* **61,** 289–294.

A'Brook, J. (1973). The effect of plant spacing on the number of aphids trapped over cocksfoot and kale crops. *Ann. Appl. Biol.* **74,** 279–285.

Adesiyun, A. A. (1978). Effects of seeding density and spatial distribution of oat plants on colonization and development of *Oscinella frit* (Diptera: Chloropidae). *J. Appl. Ecol.* **15,** 797–808.

Altieri, M. A., van Schoonhoven, A., and Doll, J. D. (1977). The ecological role of weeds in insect pest management systems: A review illustrated by bean (*Phaseolus vulgaris*) cropping systems. *PANS* **23,** 195–205.

Bach, C. E. (1980a). Effects of plant diversity and time of colonization on an herbivore–plant interaction. *Oecologia* **44,** 319–326.

Bach, C. E. (1980b). Effects of plant density and diversity on the population dynamics of a specialist herbivore: the striped cucumber beetle *Acalymma vittata*. *Ecology* **61,** 1515–1530.

Bach, C. E. (1981). Host plant growth form and diversity: Effects on abundance and feeding preference of a specialist herbivore, *Acalymma vittata* (Coleoptera: Chrysomelidae). *Oecologia* **50,** 370–375.

Belyea, H. C. (1923). The control of the white-pine weevil (*Pissodes strobi*) by mixed planting. *J. For.* **21,** 384–390.

Bennett, R. R., Turnstall, J., and Horridge, G. A. (1967). Spectral sensitivity of single retinula cells of the locust. *Z. Vgl. Physiol.* **55,** 195–206.

Blau, W. (1980). The effect of environmental disturbance on a tropical butterfly population. *Ecology* **61,** 1005–1012.

Bond, A. B. (1980). Optimal foraging in a uniform habitat: The search mechanism of the green lacewing. *Anim. Behav.* **28,** 10–19.

Bossert, W. H., and Wilson, E. O. (1963). The analysis of olfactory communication among animals. *J. Theor. Biol.* **5,** 443–469.

Brussard, P. F., and Ehrlich, P. R. (1970). The population structure of *Erebia epipsodea* (Lepidoptera: Satyrniidae). *Ecology* **51,** 119–129.

Buranday, R. P., and Raros, R. S. (1975). Effects of cabbage–tomato intercropping on the incidence and oviposition of the diamond-black moth, *Plutella xylostella* L. *Philipp. Entomol.* **2,** 369–374.

Chew, F. S. (1975). Coevolution of pierid butterflies and their cruciferous foodplants. I. The relative quality of available resources. *Oecologia* **20,** 117–12.

Chew, F. S. (1981). Coexistence and local extinction in two pierid butterflies. *Am. Nat.* **118,** 655–672.

Connell, J. H. (1978). Diversity in tropical rain forests and coral reefs. *Science* **199,** 1302–1309.

Copp, N. H., and Davenport, D. (1978). *Agraulis* and *Passiflora*. II. Behavior and sensory modalities. *Biol. Bull. (Woods Hole, Mass.)* **155,** 113–124.

Cromartie, W. J. (1975a). The effect of stand size and vegetational background on the colonization of cruciferous foodplants by herbivorous insects. *J. Appl. Ecol.* **12,** 517–533.

Cromartie, W. J. (1975b). Influence of habitat on colonization of collards by *Pieris rapae*. *Environ. Entomol.* **4,** 832–836.

Cullenward, M. J., Ehrlich, P. R., White, R. R., and Holdren, C. E. (1979). The ecology and population genetics of an alpine checkerspot butterfly, *Euphydryas anicia*. *Oecologia* **38,** 1–12.

David, W. A. L., and Gardiner, B. O. C. (1962). Oviposition and the hatching of the eggs of *Pieris brassicae* (L.) in a laboratory culture. *Bull. Entomol. Res.* **53,** 91–109.

Davis, B. N. K. (1975). The colonization of isolated patches of nettles (*Urtica dioica*) by insects. *J. Appl. Ecol.* **12,** 1–14.

Dempster, J. P. (1969). Some effects of weed control on the numbers of the small cabbage white (*Pieris rapae* L.) on brussels sprouts. *J. Appl. Ecol.* **6,** 339–346.

Dempster, J. P., and Coaker, T. H. (1974). Diversification of crop ecosystems as a means of controlling pests. *Proc. 13th Symp. Br. Ecol. Soc.* **13,** 106–114.

den Boer, P. J. (1970). On the significance of dispersal power for populations of Carabid beetles (Coleoptera, Carabidae). *Oecologia* **4,** 1–28.

Denno, R. F., Raupp, M. J., Tallamy, D. W., and Reichelderfer, C. F. (1980). Migration in heterogeneous environments: Differences in habitat selection between the wing forms of the dimorphic planthopper *Prokelisia marginata*. *Ecology* **61,** 859–867.

Dethier, V. G. (1959). Foodplant distribution and density and larval dispersal as factors affecting insect populations. *Can. Entomol.* **91,** 581–596.

Dethier, V. G. (1980). Evolution of receptor sensitivity to secondary plant substances with special reference to deterrents. *Amer. Natur.* **115,** 45–66.

Douwes, P. (1968). Host selection and host finding in the egg-laying female *Cidaria albulata* L. (Lep.: Geometridae). *Opusc. Entomol.* **33,** 233–279.

Ehrlich, P., and Raven, P. (1964). Butterflies and plants: A study in coevolution. *Evolution (Lawrence, Kans.)* **18,** 586–608.

Gibson, I. A. S., and Jones, T. (1977). Monocultures as the origins of major forest pests and diseases. *Proc. Symp. Br. Ecol. Soc.* **18,** 139–162.

Golik, Z., and Pienkowski, R. L. (1969). The influence of temperature on host orientation by the alfalfa weevil, *Hypera postica*. *Entomol. Exp. Appl.* **12,** 133–138.

Hamilton, R. J., Munro, J., and Rowe, J. M. (1978). Continuous rearing of *Oscinella frit* L. and the interaction of *O. frit* with *Avena sativa*. *Entomol. Exp. Appl.* **24,** 383–386.

Harper, J. L. (1977). "Population Biology of Plants." Academic Press, New York.

Haskell, P. T., Paskin, M. W. J., and Moorhouse, J. E. (1962). Laboratory observations on factors affecting the movements of hoppers of the desert locust. *J. Insect. Physiol.* **8,** 53–78.

Hawkes, C. (1974). Dispersal of adult cabbage rootfly (*Erioischia brassicae*) in relation to a brassica crop. *J. Appl. Ecol.* **11,** 83–93.

Hawkes, C., Patton, S., and Coaker, T. H. (1978). Mechanisms of hostplant finding in adult cabbage root fly, *Delia brassicae*. *Entomol. Exp. Appl.* **24,** 419–427.

Hayes, J. L. (1981). The population ecology of a natural population of the pierid butterfly *Colias alexandra*. *Oecologia* **49,** 188–200.

Horn, D. J. (1981). Effect of weedy backgrounds on colonization of collards by green peach aphid (*Myzus persicae*), and its major predators. *Environ. Entomol.* **10,** 285–289.

Huber, F. (1978). The insect nervous system and insect behavior. *Anim. Behav.* **26,** 969–981.

Janzen, D. H. (1970). Herbivores and the number of tree species in tropical rain forests. *Am. Nat.* **104,** 501–528.

Janzen, D. H. (1971). *Cassia grandis* L. beans and their escape from predators: A study in tropical predator satiation. *Ecology* **52,** 964–979.

Janzen, D. H. (1972). Escape in space of *Sterculia apetala* seeds from the bug *Dysdercus fasciatus* in a Costa Rican deciduous forest. *Ecology* **53,** 350–361.

Jones, R. E. (1977). Movement patterns and egg distribution in cabbage butterflies. *J. Anim. Ecol.* **46,** 195–212.

Kareiva, P. (1982). Experimental and mathematical analyses of herbivore movements: Quantifying the influence of plant spacing and quality on foraging discrimination. *Ecol. Monogr.,* **52,** 261–282.

Kareiva, P. (1983). The impact of plant community texture on herbivore populations. *In* "The Impact of Variable Host Quality on Herbivorous Insects" (R. F. Denno, and M. McClure, eds.). Academic Press, New York, in press.

Kennedy, J. S., and Fosbrooke, I. H. M. (1973). The plant in the life of an aphid. *Symp. R. Entomol. Soc. London* **6,** 129–140.

Kennedy, J. S., and Moorehouse, J. E. (1969). Laboratory observations on locust responses to wind-borne grass odour. *Entomol. Exp. Appl.* **12,** 487–503.

Kennedy, J. S., Booth, C. O., and Kershaw, W. J. S. (1961). Host finding by aphids in the field. III. Visual attraction. *Ann. Appl. Biol.* **49,** 1–21.

Kershaw, A. W. (1975). "Quantitative and Dynamic Plant Ecology." Clowes, London.

Kretzschmar, G. P. (1948). Soybean insects in Minnesota with special reference to sampling techniques. *J. Econ. Entomol.* **41,** 586–591.

Lawrence, W. S. (1983). Between and within patch movement of *Tetraopes* (Coleoptera: Cerambycidae). *Oecologia,* in press.

Lewis, C. T., and Waloff, N. (1964). The use of radioactive tracers in the study of dispersion of *Orthotylus virescens* (Douglas and Scott) (Miridae: Heteroptera). *Entomol. Exp. Appl.* **7,** 15–24.

Lower, R. L. (1972). Effect of surrounding cultivar when screening cucumber for resistance to cucumber beetle and pickleworm. *J. Am. Soc. Hortic. Sci.* **97,** 616–618.

Luginbill, P., and McNeal, F. H. (1958). Influence of seedling density and row spacings on the resistance of spring wheats to the wheat stem sawfly. *J. Econ. Entomol.* **51,** 804–808.

Mayse, M. A. (1978). Effects of spacing between rows on soybean arthropod populations. *J. Appl. Ecol.* **15,** 439–450.

McKibben, G. H., Mitchell, E. B., Scott, W. P., and Hedin, P. A. (1977). Boll weevils are attracted to volatile oils from cotton plants. *Environ. Entomol.* **6,** 804–806.

McLain, D. K. (1981). Numerical responses of *Murgantia histrionica* to concentrations of its host plant. *J. Georgia Entomol. Soc.* **16,** 257–260.

Milthorpe, F. L., and Moorby, J. (1974). "An Introduction to Crop Physiology." Cambridge Univ. Press, London.

Moericke, V. (1969). Hostplant specific color behaviour by *Hyalopterous pruni* (Aphididae). *Entomol. Exp. Appl.* **12,** 524–534.

Moericke, V., Prokopy, R. J., Berlocher, S., and Bush, G. L. (1975). Visual stimuli eliciting attraction of *Rhagoletis pomonella* (Diptera: Tephritidae) flies to trees. *Entomol. Exp. Appl.* **18,** 497–507.

Monteith, L. G. (1960). Influence of plants other than the foodplants of their host on host-finding abilities by tachinid parasites. *Can. Entomol.* **92,** 641–652.

Mowat, D. J. (1974). Factors affecting the abundance of shoot-flies (Diptera) in grassland. *J. Appl. Ecol.* **11,** 951–962.

Murdie, G., and Hassell, M. P. (1973). Food distribution, searching success, and predator-prey models. *In* "The Mathematical Theory of the Dynamics of Biological Populations" (M. S. Bartlett, and R. W. Hiorns, eds.), pp. 87–101. Academic Press, New York and London.

Oaten, A. A. (1977). Transit time and density-dependent predation on a patchily-distributed prey. *Amer. Natur.* **111,** 1061–1075.

Parencia, C. R., Davis, J. W., and Cowan, C. B. (1964). Studies on the ability of overwintered boll weevils to find fruiting cotton plants. *J. Econ. Entomol.* **57,** 162.

Perrin, R. M. (1977). Pest management in multiple cropping systems. *Agro–Ecosystems* **3,** 93–118.

Perrin, R. M., and Phillips, M. L. (1978). Some effects of mixed cropping on the population dynamics of insect pests. *Entomol. Exp. Appl.* **24,** 585–593.

Phillips, M. L. (1977). "Some Effects of Inter-Cropping Brussels Sprouts and Tomatoes on Infestations of *P. maculipennis* (Curt.) and *Aleyrodes brassicae* (Walk.)." M.S. Thesis, Univ. of London.

Pimentel, D. (1961a). The influence of plant spatial patterns on insect populations. *Ann. Entomol. Soc. Am.* **54,** 61–69.

Pimentel, D. (1961b). Species diversity and insect population outbreaks. *Ann. Entomol. Soc. Am.* **54,** 76–86.

Price, P. (1976). Colonization of crops by arthropods: Non-equilibrium communities in soybean fields. *Environ. Entomol.* **5,** 605–611.

Prokopy, R. J., and Owens, E. D. (1978). Visual generalist with visual specialist phytophagous insects: Host selection behavior and application to management. *Entomol. Exp. Appl.* **24,** 609–620.

Prokopy, R. J., Moericke, V., and Bush, G. L. (1973). Attraction of apple maggot flies to the odor of apple. *Environ. Entomol.* **2,** 743–750.

Ralph, C. P. (1976). Search behavior of the large milkweed bug, *Oncopeltus fasciatus* (Hemiptera: Lygaeidae). *Ann. Entomol. Soc. Am.* **70,** 337–342.

Ralph, C. P. (1977). Effects of host plant density on populations of a specialized seed-sucking bug, *Oncopeltus fasciatus*. *Ecology* **58,** 799–809.

Rausher, M. D. (1979). Egg recognition: Its advantage to a butterfly. *Anim. Behav.* **27,** 1034–1040.

Rausher, M. D. (1981). The effect of native vegetation on the susceptibility of *Aristolochia reticulata* (Aristolochiaceae) to herbivore attack. *Ecology* **62,** 1187–1195.

Read, D. P., Feeny, P. P., and Root, R. B. (1970). Habitat selection by the aphid parasite *Diaeretiella rapae* and hyperparasite *Charips brassicae*. *Can. Entomol.* **102,** 1567–1578.

Risch, S. J. (1980). The population dynamics of several herbivorous beetles in a tropical agroecosystem: The effect of interplanting corn, beans, and squash in Costa Rica. *J. Appl. Ecol.* **17,** 593–612.

Risch, S. J. (1981). Insect herbivore abundance in tropical monocultures and polycultures: An experimental test of two hypotheses. *Ecology* **62,** 1325–1340.

Robert, P. C., and Blaisinger, P. (1978). Role of non-host plant chemicals in the reproduction of an oligophagous insect: The sugar beet moth *Scrobipalpa ocellatella* (Lepidoptera: Gelechidae). *Entomol. Exp. Appl.* **24,** 632–636.

Rogers, D. (1972). Random search and insect population models. *J. Anim. Ecol.* **41,** 369–383.

Rogers, D., and Hubbard, S. (1974). How the behaviour of parasites and predators promotes population stability. *In* "Ecological Stability" (M. B. Usher and M. H. Williamson, eds.), pp. 99–121. Chapman and Hall, Oxford.

Root, R. B. (1973). Organization of plant-arthropod associations in simple and diverse habitats: The fauna of collards (*Brassica oleracea*). *Ecol. Monogr.* **43,** 95–124.

Root, R. B. (1975). Some consequences of ecosystem texture. *In* "Ecosystem Analysis and Prediction" (S. A. Levin, ed.), pp. 83–97. Soc. Industrial & Appl. Maths., Philadelphia, Pennsylvania.

Ryan, J., Ryan, M. F., and McNaeidhe, F. (1980). The effect of interrow plant cover on populations of the cabbage root fly *Delia brassicae* (Wiedemann). *J. Appl. Ecol.* **17,** 31–40.

Sauer, D., and Feir, D. (1973). Studies on natural populations of *Oncopeltus fasciatus* (Dallas), the large milkweed bug. *Am. Midl. Nat.* **90,** 13–37.

Saxena, K. N., and Goyal, S. (1978). Host-plant relations of the citrus butterfly *Papilio demoleus* L.: Orientational and ovipositional responses. *Entomol. Exp. Appl.* **24,** 1–10.

Saxena, K. N., and Prabha, S. (1977). Relationship between the olfactory sensilla of *Papilio demoleus* L. larvae and their orientation responses to different odours. *J. Entomol. Ser A* **50,** 119–126.

Saxena, K. N., and Saxena, R. C. (1975). Patterns of relationships between certain leafhoppers and plants. III. Range and interaction of sensory stimuli. *Entomol. Exp. Appl.* **18,** 194–206.

Schneider, D. (1974). The sex-attractant receptor of moths. *Sci. Am.* **231,** 28–35.

Schoonhoven, L. M. (1973). Plant recognition by lepidopterous larvae. *In Symp. R. Entomol. Soc. London* **6,** 87–99.

Scriber, J. M. (1977). Limiting effects of low leaf-water content on the nitrogen utilization, energy budget, and larval growth of *Hyalophora cecropia* (Lepidoptera: Saturniidae). *Oecologia* **28,** 269–287.

Simmons, G. A., Leonard, D. E., and Chen, C. W. (1975). Influence of tree species density and composition of parasitism of the spruce budworm, *Choristoneura fumiferana* (Clem.). *Environ. Entomol.* **4,** 832–836.

Singer, M. C. (1971). Evolution of food-plant preference in the butterfly *Euphydryas editha. Evolution (Lawrence, Kans.)* **25,** 383–389.

Sloderbeck, P. E., and Edwards, C. R. (1979). Effects of soybean cropping practices on mexican bean beetle and red-legged grasshopper populations. *J. Econ. Entomol.* **72,** 850–853.

Smith, J. G. (1969). Some effects of crop background on populations of aphids and their natural enemies on Brussels sprouts. *Ann. Appl. Biol.* **63,** 326–330.

Smith, J. G. (1976). Influence of crop background on aphids and other phytophagous insects on brussels sprouts. *Ann. Appl. Biol.* **83,** 1–13.

Smith, R. W., and Whittaker, J. B. (1980). The influence of habitat type on the population dynamics of *Gastrophysa viridula* Degeer (Coleoptera: Chrysomelidae). *J. Anim. Ecol.* **49,** 225–236.

Solbreck, C., and Pehrson, I. (1979). Relations between environment, migration and reproduction in a seed bug *Neacoryphus bicrocis* (Heteroptera: Lygaeidae). *Oecologia* **43,** 51–62.

Southwood, T. R. E. (1962). Migration of terrestrial arthropods in relation to habitat. *Biol. Rev. Cambridge Philos. Soc.* **37,** 171–214.

Staedler, E. (1976). Sensory aspects of insect plant interactions. *Proc. Int. Congr. Entomol. 15th,* pp. 228–248.

Stanton, M. L. (1982). Searching in a patchy environment: Foodplant selection by *Colias p. eriphyle* butterflies. *Ecology,* **63,** 839–853.

Strong, D. R. (1979). Biogeographic dynamics of insect–hostplant communities. *Annu. Rev. Entomol.* **24,** 89–119.

Sutton, O. G. (1949). Diffusion of gasses in the lower atmosphere. *Q. J. R. Meteorol. Soc.* **73,** 257–280.

Tabashnik, B. E. (1980). Population structure of pierid butterflies. III. Pest populations of *Colias Philodice eriphyle*. *Oecologia* **47,** 175–183.

Tahvanainen, J. O., and Root, R. B. (1972). The influence of vegetational diversity on the population ecology of a specialized herbivore, *Phyllotreta cruciferae* (Coleoptera: Chrysomelidae). *Oecologia* **10,** 321–346.

Theunissen, J., and Den Ouden, H. (1980). Effects of intercropping with *Spergula arvensis* on pests of Brussels sprouts. *Entomol. Exp. Appl.* **27,** 260–268.

Thompson, J. N., and Price, P. W. (1977). Plant plasticity, phenology, and herbivore dispersion: Wild parsnip and the parsnip webworm. *Ecology* **58,** 1112–1119.

Thorsteinson, A. J. (1960). Host selection in phytophagous insects. *Annu. Rev. Entomol.* **5,** 193–218.

Traynier, R. M. M. (1967). Effect of host plant odor on the behaviour of the adult cabbage root fly, *Erioschia brassicae. Entomol. Exp. Appl.* **10,** 321–328.

Turner, N. C., and Friend, R. B. (1933). Cultural practices in relation to Mexican bean beetle control. *J. Econ. Entomol.* **26,** 115–123.

Vandermeer, J. H. (1974). Relative isolation and seed predation in *Calliandra grandiflora,* a Mimosaceous legume from the highlands of Guatemala. *Biotropica* **6,** 267–268.

van Emden, H. F. (1965). The effect of uncultivated land on the distribution of the cabbage aphid (*Brevicoryne brassicae*) on an adjacent crop. *J. Appl. Ecol.* **2,** 171–196.

van Emden, H. F., and Williams, G. F. (1974). Insect stability and diversity in agro-ecosystems. *Annu. Rev. Entomol.* **19,** 455–475.

Vinson, S. B. (1981). Habitat location. *In* "Semiochemicals: Their Role in Pest Control" (D. A. Norlund, R. L. Jones, and W. J. Lewis, eds.), pp. 51–77. Wiley, New York.

Visser, J. H., and Nielson, J. K. (1977). Specificity in the olfactory orientation of the Colorado beetle *Leptinotarsa decemlineata. Entomol. Exp. Appl.* **21,** 14–22.

Wallace, G. K. (1958). Some experiments on form perception in the nymphs of the desert locust *Schistocerca gregaria* F. *J. Exp. Biol.* **35,** 765–775.

Watt, W. B., Han, D., and Tabashnik, B. E. (1979). Population structure of Pierid butterflies. II. A "native" population of *Colias philodice eriphyle* in Colorado. *Oecologia* **44,** 44–52.

Way, M. J., and Heathcote, G. D. (1966). Interactions of crop density of field beans, abundance of *Aphis fabae* Scop., virus incidence, and aphid control by chemicals. *Ann. Appl. Biol.* **57,** 409–423.

Wehner, R. (1975). Pattern recognition in insects. *In* "The compound eye and vision of insects" (G. A. Horridge, ed.), pp. 75–112. Clarendon Press, Oxford.

Wood, D. L. (1973). Selection and colonization of Ponderosa pine by bark beetles. *Symp. R. Entomol. Soc. London* **6,** 101–117.

Wright, R. H. (1958). The olfactory guidance of flying insects. *Can. Entomol.* **90,** 81–89.

# PART III

# Host Search in Relation to the Breadth of Diet

# 5

# Integration of Visual Stimuli, Host Odorants, and Pheromones by Bark Beetles and Weevils in Locating and Colonizing Host Trees

GERALD N. LANIER



## I. INTRODUCTION

Bark beetles (Scolytidae) and bark weevils (certain Curculionidae) utilize the phloem–cambium region of trees as breeding substrate. Tunneling in the phloem is incompatible with the life functions of this vital tissue. Conversely, breeding cannot succeed if the phloem remains alive because host antibiosis will overcome the invading insects; the infested tree, or part of the tree, must die if reproduction of the insect is to succeed.

The phloem–cambium region of a tree is a temporary habitat in that only one brood can be produced in a given unit of host; each generation must find a new place to breed. Opportunities for breeding are maximized by dispersal of the emerging population, finding of new hosts by visual and olfactory cues, acceptance of the host, and finally the release of odorants that concentrate the extant population. The concentration phase ensures efficient utilization of a scarce



HERBIVOROUS INSECTS

ISBN 0-12-045580-3

resource and the generation of a mass attack that may overcome resistance of the host and make it available for breeding.

Many bark beetles and weevil adults undergo distinct feeding and breeding phases. A feeding phase may be necessary to sustain the insect through reproductive diapause or during dispersal until a breeding habitat is found. Breeding and feeding hosts are usually the same species but they are in different physiological conditions. The host used for feeding may be too vigorous for a successful breeding attack but some feeding always occurs on the breeding host.

Some species breed only in recently dead or moribund host material. Other species must kill vital hosts in order to perpetuate the population. The majority of the species occupy an intermediate zone between the saprophage and predator: they are opportunists that normally colonize trees weakened by age, mechanical factors, defoliation, or drought, but they are capable of killing healthy trees when the population explodes following an increase in available breeding material by windthrow, fire, or logging. Once the population passes a threshold at which host resistance (antibiosis) is a limiting factor, a devastating outbreak may virtually eliminate the host species within local areas. The enormous economic losses and profound ecological changes perpetrated by bark beetles have stimulated extensive study of their biology and control.

In the saprophagous mode, bark beetles and weevils face the typical problems of locating a breeding substrate that is often scarce and discontinuous in time and space. In the tree-killing mode, they are small predators that must overcome large prey. Their ability to locate and maximize utilization of susceptible trees effectively or to mass attack to overcome the resistance of vigorous hosts is dependent on volatile chemical signals produced by the host (kairomones) and other beetles (pheromones). The release of pheromones is a signal that a host has been located (by vision or odor) and accepted (by gustation and successful penetration).

Reviewed and contrasted in this chapter are host selection and colonization by three insects within the saprophage–predator continuum: an obligatory saprophage, the native elm bark beetle [*Hylurgopinus rufipes* (Eich)]; a facultative predator–saprophage, the European elm bark beetle [*Scolytus multistriatus* (Marsh)]; and a predator, the white pine weevil [*Pissodes strobi* (Peck)].

## II. HOST SELECTION BY THREE SPECIES

### A. *Hylurgopinus rufipes*

The native elm bark beetle breeds in moribund elm, and feeds and overwinters in bark of healthy elm trees. Reports (Kaston, 1939; Baker, 1972) that this insect also utilizes cherry (*Prunus*), basswood (*Tilia*), and ash (*Fraxinis*) as breeding material required scrutinization because the engravings of *Leperisinus aculeatus* (Say), which attacks ash, and *Phleotribus liminaris* (Harris), which breeds in cherry, are similar to the work of *Hylurgopinus rufipes*. There are no scolytid galleries in basswood that resemble those of *H. rufipes*, but this tree might be confused with elm because of similarities in bark, size, and growth habit.

To test the validity of the dubious host records, we confined *H. rufipes* (mixed sexes) on small logs of fire cherry (*Prunus pennsylvanica* L.), white ash (*Fraxinis americana* L.), basswood (*Tilia glabra* Vent.), and American elm (*Ulmus americana* L.). Beetles colonized the elm logs, but made only a few bite marks on logs of the other species; rather than penetrating the bark of ash, basswood, or cherry, the beetles perished in the container. The questionable host attributions were further tested in the field by establishing separate log piles of each of the tree species and of all of the species combined. Most of the ash logs were attacked by *Leperisinus aculeatus* whereas some of the cherry logs were colonized by *Phleotribus liminaris*. *Hylurgopinus rufipes* was found only in elm logs, and no bark beetle species utilized the basswood. Log piles of mixed species must have produced a melange of odors through which the beetles navigated in order to find their respective hosts.

One afternoon during late May 1981, *H. rufipes* and *L. aculeatus* were observed arriving at one of the log piles. No account of the landing rates was made but it seemed clear that the majority of the beetles circled over and landed on logs of their respective host species. Individuals that landed on a substrate other than a host log usually launched into flight after a brief pause that included no apparent biting of the bark. Beetles that landed on a host log generally remained stationary for a few seconds, then walked over the bark until they settled into crevices or located and attempted to enter tunnels that were already established. Both species appeared to be guided by kairomones and/or pheromones coming from the infested logs, and both appeared to be able to distinguish between host and nonhost logs without biting them.

From these experiments and observations it is apparent that *H. rufipes* is restricted to *Ulmus* species (monophagous in the broad sense), that it locates its host by odorants, and that it can identify elm logs in a condition suitable for breeding prior to penetrating the bark.

*Hylurgopinus rufipes* that fly from overwintering sights in the spring are very sensitive to odorants from moribund elms. Gardiner (1979) showed that elms killed with the herbicide cacodylic acid were very attractive to *H. rufipes*. We found that the period of attractiveness of cacodylic acid-treated trees extended before and after the period during which the beetles penetrated the bark to establish breeding galleries. Apparently the emission of attractants initially occurred before the trees were moribund enough that the beetles would attempt to bore into them and it continued beyond the point at which the beetles would reject that host for further colonization.

It seemed probable that mass colonization by *Hylurgopinus rufipes* would be stimulated by pheromones as is the case for many Scolytidae, including the European elm bark beetle, *S. multistriatus*. However, repeated field tests have failed to demonstrate that a beetle-produced odorant is essential to the aggregation of *H. rufipes* on breeding material. Uninfested logs were as attractive to both sexes of the native elm bark beetle as logs in which males, females, or both sexes were boring (D. P. O'Callaghan, unpublished report, 1979). Gardiner (1979) found that an ethanol extract of the bark of an elm that was treated with cacodylic acid attracted pedestrian *H. rufipes* in a laboratory bioassay. We found similar extract and an ethanol control to be minimally attractive in the field, but an extract of volatiles (Porapak extract method of Byrne *et al.*, 1975) from entrained air passing over moribund elm logs appeared to be strongly attractive to flying *H. rufipes*.

These results indicate that the attraction of *H. rufipes* to breeding sites is based on compounds produced in deteriorating elm tissue. The compound or the bouquet is unique enough to identify the plant (elm) and its condition (moribund). The period of production of the host attractant exceeds the period during which the host is actually accepted for colonization. Penetration of the bark may be governed by factors such as phloem moisture or the presence of beetles already utilizing the available breeding space. Pheromones may be involved in sex recognition and spacing of the attacks, but they are not responsible for long-range attraction to the breeding host.

Progeny of the spring-flying *H. rufipes* emerge in the late summer and fall. Despite their abundance, few beetles are attracted to moribund elms and very few initiate breeding galleries. Sticky bands placed on elms, other tree species, and utility poles indicate that the beetles disperse widely and land on many vertical objects (D. P. O'Callaghan, unpublished report, 1979). Whether they land more frequently on elms than on other trees has not been determined. By whatever means the healthy elm is located, a large portion of the population finds an overwintering host by late fall. Once the beetle has landed, it is clear that it is able to recognize an elm in which to bore feeding niches. As winter approaches, the beetles move from branches to the base of the tree. The possibility that beetles find a healthy elm by trial and error, then release a pheromone to attract others of its kind to the tree, has not been investigated.

### B. *Scolytus multistriatus*

*Scolytus multistriatus*, the European elm bark beetle, also restricts its feeding and breeding to elms (*Ulmus*). It frequently attacks and sometimes kills apparently healthy trees or trees that are only partly infected by the Dutch elm disease (DED) fungus *Ceratocystus ulmi*. Colonization begins in the diseased limbs and quickly spreads to undiseased areas. Rapid death of the tree, usually attributed to

DED, actually results from successful colonization by the beetles. The usual mass attack on elm wood by *S. multistriatus* before it becomes sufficiently moribund to be acceptable for *H. rufipes* accounts for the virtual elimination of the native species wherever winter temperatures are not low enough to affect survival of *S. multistriatus* larvae in the outer bark.

Adults emerging from the brood host make a dispersal flight during which they are relatively unresponsive to odorants (Lanier *et al.*, 1976). After a flight that is sufficient (about 400 m) to overcome the suppression of olfactory stimulation, beetles are attracted by volatiles from elm wood that is stressed by disease, drought, or injury. Virgin females boring into bark produce an aggregation pheromone that induces mass attraction, breeding, and colonization. The pheromone, which attracts both sexes, consists of 4-methyl-3-heptanol (H) and (−)-α-multistriatin (M) released by the female plus (−)-α-cubebene (C) produced in the tree (Gore *et al.*, 1977). All three compounds are necessary for maximum attractiveness so the host must be considered to be an essential part of the signal. Presumably, potential hosts that did not produce C when attacked could not be extensively colonized. After mating, females terminate the release of H but continue to release M. When M is in excess of H (most of the females are mated), colonization is terminated.

*Scolytus multistriatus* that do not locate a breeding host after their initial dispersal flight can sustain themselves by feeding in twig crotches of healthy elms. Whether the feeding host is initially located by olfaction, vision, or chance is not known.

Twig feeding is highly contagious in distribution between and within trees. This suggests that beetles that locate a feeding host signal their conspecifics of its location by pheromones. Our laboratory bioassays showed that the attractiveness of freshly cut elm twigs was enhanced by beetles in them. The fact that mating in twig crotches is common (Bartels and Lanier, 1974; Svihra and Clark, 1980) also suggests that *S. multistriatus* may use a pheromone system during the feeding and during the colonization phases.

We first investigated the possible role of the aggregation pheromone components on twig feeding because trapping experiments showed that increasing M relative to H reduced the number of beetles attracted (Cuthbert and Peacock, 1978). Such a shift in the relative quantities of the two beetle-produced compounds was shown to occur when females are mated and was thought to be responsible for the termination of attraction to brood wood that is fully occupied (Gore *et al.*, 1977). We reasoned that baiting trees with M might inhibit landing and feeding in the tree crown as it had reduced catches on sticky traps on the tree bole.

The opposite was true. M greatly increased twig crotch feeding. H alone had no effect on landing on the tree bole or on twig feeding, but when H and M were in the ratio produced by virgin females, landing on the tree bole was maximal. M

**TABLE I.** Indices of *Scolytus multistriatus* Landing on Boles and Feeding in Twig Crotches of Baited Trees in Response to Pheromone Components

| Bait[a] | Beetles Sampled on Bole | | "Prime" Crotches with Feeding | |
|---|---|---|---|---|
| | 1980 | 1981 | 1980 | 1981 |
| HMC | 100a[b] | 100a | 100a | 100ab |
| HMC + M | 20b | 46a | 148b | 140b |
| HMC + H | no test | 106a | no test | 77a |
| M | 7b | no test | 42c | no test |
| H | 0b | 0a | 6d | 23c |
| Blank | <1b | 0a | 10d | 11c |
| Base number for index | 302.3 | 36.3 | 43.14 | 20.00 |

[a]H, 4-Methyl-3-heptanol; M, α-multistriatin; C, α-cubebene. HMC signifies the three components in the natural ratio. The components H and M were added in excess and presented alone.
[b]Values are normalized as an index to HMC standard. Rates for HMC are the base numbers. The indices followed by different letters are significantly different ($p < 0.05$, Student–Newman–Kneulls test).

in excess of the virgin female ratio reduced landing on the bole and increased twig feeding (Table I) (R. J. Rabaglia and G. N. Lanier, unpublished).

Like *H. rufipes, S. multistriatus* appears to be able to sense that it is on elm wood without tasting it. Once the bark is bitten, nonvolatile feeding stimulants (Doskotch *et al.*, 1970) help sustain the penetration. Trees other than elm may lack stimulants or possess repellents such as juglone from hickories (*Carya* spp.). Baker and Norris (1968) explain that the rather high degree of specificity maintained by *S. multistriatus* in twig feeding is the combined effects of feeding stimulants in host versus their absence or their combination with feeding deterrents in nonhosts.

### C. *Pissodes strobi*

*Pissodes strobi,* the white pine weevil, breeds in terminal leaders of several species of pine and spruce. The favored host individuals are the most rapidly growing sapling and pole-sized trees in the stand. Open-grown trees are especially likely to be attacked. For this reason, *P. strobi* is usually most abundant in young plantations.

There is one generation per year. New adults emerge from the killed ("weeviled") leader in midsummer and feed on the bark of lateral branches of the host tree until fall. As temperatures drop the weevils move downward to overwinter in the litter at the base of the host from which they emerged. In spring they crawl up the brood tree or a nearby tree where they feed on the bark of lateral branches. When temperatures permit flight (about 23°C), the weevils disperse; the brood host is no longer suitable because the leader is dead and the weevils rarely accept lateral branches for breeding. When a new host is located, males release a pheromone that attracts *P. strobi* of both sexes, facilitates multiple attacks, and ensures sufficient oviposition to utilize the entire leader.

VanderSar and Borden (1977) found that pedestrian *P. strobi* oriented to vertical more than to oblique or horizontal spruce twigs, and that they approached silhouettes that suggested a terminal leader more often than other shapes. Both sexes tended to select the longest and thickest leaders. Once a potential breeding host is located, weevils reach the terminal leader by the combination of positive phototaxis and negative geotaxis (Sullivan, 1959).

Terpenes in conifer trees vary between species, within species, and within individuals. It is reasonable to hypothesize that *P. strobi* utilizes volatized terpenes to aid in the location of a potential host. Perhaps it is no coincidence that Kang (1976), with Norway spruce (*Picea abies* L.), and Wilkinson (1980), with white pine (*Pinus strobis* L.), found that host individuals with high limonene and low α-pinene content were attacked more frequently than would be predicted from the weeviling frequencies in the two populations. Laboratory assays produced a positive klinotaxis to limonene and negative responses to α-pinene and myrcene (Carlson, 1971). However, no one has published data showing that limonene or any other terpene attracted *P. strobi* in the field.

Whether or not limonene attracts in-flight *P. strobi*, it may be important in the identification of a suitable host once landing occurs. It may also aid in the identification of the terminal leader, where limonene concentration tends to be maximal during the growing season.

Limonene does not stimulate feeding, but it does synergize response to nonvolatile (and unidentified) feeding stimulants in the bark of Sitka spruce (*Picea sitchensis*) (Alfaro *et al.*, 1980). Alfaro and Borden (1982) assayed tissue of 35 conifers, 5 broad-leaved trees, and 1 fern for feeding by *Pissodes strobi* from Sitka spruce. Feeding stimulants were present in all of the conifers and absent in all of the nonconifer species. Five of the conifers and four of the nonconifers contained feeding deterrents. Surprisingly, an extract from white pine, the principal host of *P. strobi* in eastern North America, showed strong feeding deterrency to weevils from Pacific Sitka spruce. Earlier work indicated that *P. strobi* from the Pacific Coast and Rocky Mountains fed on white pine as readily as on spruce (VanderSar *et al.*, 1977).

Feeding is not synonymous with breeding. When he found that *P. strobi* from Sitka spruce would feed on western white pine as readily as eastern white pine, but not oviposit in the former species, VanderSar (1978) hypothesized that western white pine contained feeding stimulants but lacked a chemical stimulant necessary for oviposition. Indeed, western white pine planted in New England was less often weeviled than eastern white pine to which it was adjacent (Wilkinson, 1980).

Apparent host selection may in some cases be an artifact of insect survival. Phillips (1981) found that white spruce, *Picea glauca* L., was fed upon more than any of five other species of pine and spruce, including eastern white pine, from which the *Pissodes strobi* used in the assay were reared. This result was surprising because white spruce is very rarely weeviled even in plantations near heavily attacked white pine and/or Norway spruce (Table II). Weevils killed the leaders of white spruce upon which they were confined, but no brood was produced because larvae failed to mature. Other conifer species yielded brood adults when leaders were killed although the weevils appeared to have difficulty in killing leaders of red and Scots pines. White pine was the only host on which the *P. strobi* brood exceeded the number of adults confined on the leaders (Phillips, 1981).

As is the case with bark beetles, once a suitable breeding host is located and accepted, *P. strobi* apparently use mass attack as a means of overcoming host antibiosis. We have found that aggregation and mating is stimulated by male weevils on terminal leaders when they release a pheromone consisting of the terpene alcohol grandisol (first identified from the cotton boll weevil *Anthonomus grandis* Boheman; Tumlinson *et al.,* 1969) and the analogous aldehyde, grandisal (Booth *et al.,* in press). Sustained feeding is essential for pheromone release, but pheromone release occurs only during the period that coincides with shoot elongation and probably only on a potential breeding host.

**TABLE II.** *Pissodes strobi* Laboratory Feeding Preference and Field Reproductive Success upon Terminal Leaders of Six Conifer Species[a]

| | Feedings per Weevil | | | | |
|---|---|---|---|---|---|
| Conifer Species | Fall 1979 | Spring 1980 | Sample Size | Leaders Killed | Emergence per Leader |
| Eastern white pine | 0.24 | 0.86 | 5 | 5 | 25.2 |
| Red pine | 0.06 | 0.06 | 5 | 1 | 4.0 |
| Scots pine | 0.18 | 0.14 | 5 | 0 | 2.2 |
| White spruce | 1.38 | 2.50 | 4 | 3 | 0.0 |
| Norway spruce | 0.40 | 0.56 | 3 | 3 | 2.3 |
| Blue spruce | 0.86 | 0.44 | 3 | 2 | 5.0 |

[a]Phillips (1981) gives details of the test. Underlined figures significantly differ from other values in the same column ($p < 0.05$), Student–Newman–Kneulls test.

## III. GENERALIZATIONS AND SUMMARY

Bark beetles and bark weevils breed in the phloem–cambium region of woody plants. This always accompanies death and degradation of the tissue; thus the host becomes unsuitable for further breeding. To find a new breeding site, beetles must undertake hazardous dispersal flights. Locating a new host is aided by volatile components from the host and by pheromones. Landing and short-range orientation is guided by vision.

The relative importance of host volatiles and pheromones is generally related to the condition of the host when it is attacked. *Hylurgopinus rufipes* locates the degradating hosts in which it breeds primarily, and perhaps entirely, by compounds emitted from moribund elm tissue. *Scolytus multistriatus,* which attacks elms in an early state of decline, also initially utilizes host odorants to locate breeding material, but the mass attack and killing of this material is orchestrated by a pheromone consisting of three compounds—two released by the virgin females boring into the bark and the third produced by the tree in response to the invasion. There is no evidence that *Pissodes strobi* locates a breeding site by orienting to host volatiles; the healthy trees that this weevil attacks are probably not sufficiently distinct for this to be an efficient host-locating mechanism. *P. strobi* might be attracted by host volatiles to the general area where the host grows, but selection of a specific individual is probably guided by orientation to the vertical silhouette of a terminal leader. Once a suitable host is located, the male releases an aggregation pheromone that induces attack by a number of weevils sufficient to kill and completely colonize the leader.

In addition to forming the aggregations necessary to kill a resistant host, pheromones may govern attack density, mating, and the efficient utilization of the available host. In the case of *S. multistriatus,* a change in the pheromone that accompanies mating shifts the attack to another breeding host or to twig crotches where they feed.

Each of the insects detailed herein may feed on hosts that are in a condition that makes them unavailable for breeding. Pheromones are a factor in feeding attack of *Scolytus multistriatus* and possibly *Hylurgopinus rufipes,* but they appear not to be a factor in feeding of *Pissodes strobi* other than that which occurs on the breeding host.

Most bark beetles and bark weevils specialize on a few host species of the same genus. Some are limited to a single host species. A few utilize hosts of different genera, but almost none use trees of different families. Host selection is especially critical for subcortical insects because they must exist within an atmosphere of defensive chemicals that differ among tree species. Selecting the proper host is a matter of surviving the chemical environment, of finding adequate nutrition, and sometimes of obtaining a precursor or synergist for its aggregation pheromone.

## REFERENCES

Alfaro, R. I., and Borden, J. H. (1982). Host selection by the white pine weevil, *Pissodes strobi* Peck: Feeding bioassays using host and nonhost plants. *Can. J. For. Res.* **12,** 64–70.

Alfaro, R. I., Pierce, H. D., Borden, J. H., and Oehlschlager, A. C. (1980). Role of volatile and nonvolatile components of Sitka spruce bark as feeding stimulants for *Pissodes strobi* Peck (Coleoptera: Curculionidae). *Can. J. Zool.* **58,** 626–632.

Baker, J. E., and Norris, D. M. (1968). Behavioral responses of the smaller European elm bark beetle, *Scolytus multistriatus* to extracts of non-host tree tissues. *Entomol. Exp. Appl.* **11,** 464–469.

Baker, W. L. (1972). "Eastern Forest Insects," USDA Misc. Publ. 1175. US Govt. Printing Office, Washington, D.C.

Bartels, J. M., and Lanier, G. N. (1974). Emergence and mating in *Scolytus multistriatus* (Coleoptera: Scolytidae). *Ann. Entomol. Soc. Am.* **67,** 365–370.

Booth, D. C., Phillips, T. W., Claesson, A., Silverstein, R. M., Lanier, G. N., and West, J. R. (1982). Aggregation pheromone components of two species of *Pissodes* weevils (Coleoptera: Scolytidae): Isolation, identification and field activity. *J. Chem. Ecol.,* **9,** 1–12.

Byrne, K. J., Gore, W. E., Pearce, G. T., and Silverstein, R. M. (1975). Porapak-Q collection of airborne organic compounds serving as models for insect pheromones. *J. Chem. Ecol.* **1,** 1–7.

Carlson, R. L. (1971). "Behavior of the Sitka Spruce Weevil, *Pissodes Stichensis* Hopkins (Coleoptera: Curculionidae), in Southern Washington." Ph.D. Dissertation, Univ. of Washington, Seattle.

Cuthbert, R. A., and Peacock, J. W. (1978). Response of the elm bark beetle, *Scolytus multistriatus* (Coleoptera: Scolytidae), to component mixtures and doses of the pheromone, Multilure. *J. Chem. Ecol.* **4,** 363–373.

Doskotch, R. W., Chatlerji, S. K., and Peacock, J. W. (1970). Elm bark derived feeding stimulants for the smaller European elm bark beetle. *Science (Washington, D.C.)* **167,** 380–382.

Gardiner, L. M. (1979). Attraction of *Hylurgopinus rufipes* to cacodylic acid–treated elms. *Bull. Entomol. Soc. Am.* **25,** 102–104.

Gore, W. E., Pearce, G. T., Lanier, G. N., Simeone, J. B., Silverstein, R. M., Peacock, J. W., and Cuthbert, R. A. (1977). Aggregation attractant of the European elm bark beetle, *Scolytus multistriatus:* Production of individual components and related aggregation behavior. *J. Chem. Ecol.* **3,** 429–446.

Kang, H. C. (1976). "Correlation between White Pine Weevil Damage and Tree Height and Monoterpenes in Norway Spruce Plantations." M.Sc. Thesis, State Univ. of New York Coll. Environ. Sci. and Forestry, Syracuse.

Kaston, B. J. (1939). The native elm bark beetle, *Hylurgopinus rufipes* (Eichhoff) in Connecticut. *Bull. Conn. Agric. Exp. Stn. New Haven,* No. 420.

Lanier, G. N., Silverstein, R. M., and Peacock, J. W. (1976). Attractant pheromone of the European elm bark beetle (*Scolytus multistriatus*): Isolation, identification, synthesis and utilization studies. *In* "Perspectives in Forest Entomology" (J. E. Anderson, and H. K. Kaya, eds.), pp. 149–175. Academic Press, New York.

Phillips, T. W. (1981). "Aspects of Host Preference and Chemically Mediated Aggregation in *Pissodes strobi* Peck and *P. approximatus* Hopkins (Coleoptera: Curculionidae)." M.Sc. Thesis, State Univ. of New York Coll. Environ. Sci. and Forestry, Syracuse.

Sullivan, C. P. (1959). The effect of light and temperature on the behavior of adults of the white pine weevil, Pissodes strobi (Peck). *Can. Entomol.* **91,** 213–232.

Svihra, P., and Clark, J. K. (1980). The courtship of the elm bark beetle. *Calif. Agric.* **April,** 7–9.

Tumlinsen, J. H., Hardee, D. D., Gueldner, R. C., Thompson, A. C., Hedin, P. A., and Minyard, P. (1969). Sex pheromone produced by male bole weevils: Isolation, identification, and synthesis. *Science* **166,** 1010–1012.

VanderSar, T. J. D. (1978). Resistance of white pine to feeding and oviposition by *Pissodes strobi* Peck in western Canada. *J. Chem. Ecol.* **4,** 641–647.

VanderSar, T. J. D., and Borden, J. H. (1977). Visual orientation of *Pissodes strobi* Peck (Coleoptera: Curculionidae) in relation to host selection behaviour. *Can. J. Zool.* **55,** 2042–2049.

VanderSar, T. J. D., Borden, J. H., and McLean, J. A. (1977). Host preference of *Pissodes strobi* Peck (Coleoptera: Curculionidae) reared from three native hosts. *J. Chem. Ecol.* **3,** 377–389.

Wilkinson, R. C. (1980). Relationship between cortical monoterpenes and susceptibility of eastern white pine to white pine weevil attack. *For. Sci.* **26,** 581–589.

# 6

# Host Location in the Colorado Potato Beetle: Searching Mechanisms in Relation to Oligophagy*

MICHAEL L. MAY AND SAMI AHMAD



## I. INTRODUCTION

A recurrent theme of this book is the relation of breadth of feeding niche to host-plant finding by insects. The aim in this chapter is to examine the behaviors by which certain oligophagous insects locate and select food plants. Particular emphasis will be placed on the Colorado potato beetle *Leptinotarsa decemlineata*

*Preparation of this manuscript was supported in part by Hatch Act Funds under the New Jersey Agricultural Experiment Station Projects 08149 and 08130. Publication No. F-08149: 08130-02-82.



HERBIVOROUS INSECTS

ISBN 0-12-045580-3

(Say). This insect is of special interest because of its great economic importance, because extensive research has focused on it, and because the results of that research may represent an important advance in our understanding of the use by insects of plant chemistry as a cue for host location.

One of the many difficulties in undertaking such a review is the lack of agreement as to what constitutes oligophagy, monophagy, or polyphagy. Table I gives a sample sufficient to illustrate the divergence of opinion on this subject. In part this is the result of attempting to break a nearly continuous spectrum of response into discrete categories that must remain largely arbitrary. Additional problems arise in assigning specific insects to certain categories. In some cases this is merely the result of insufficient knowledge of the host range, but in others it may reflect genuine changes in host range, either over time or in space. The recent report of Fox and Morrow (1981) has shown that even species that are quite generalized in their diet over their entire geographic range may be locally far more specialized. This kind of local specialization may be a common property of generalized herbivores; the latter may then be faced with problems of host location and selection that are very similar to monophagous or oligophagous species throughout their range. Also, characteristics of the physical environment may affect host selection by individual insects (de Wilde *et al.,* 1969a; Saxena, 1978).

It is clear, then, that as Dethier (1978) suggested, the characteristics of food specialists and food generalists represent "variations on a theme rather than unique themes [and are] incredibly more complex then we had in our innocence imagined [p. 760]." Thus, regardless of our definitions of feeding categories, we recognize that understanding of the feeding strategies of any insect may be enhanced by understanding of and comparison with other groups that display both similar and contrasting strategies.

In this chapter the primary consideration is of insects that appear to be obligate oligophages throughout their range. The Colorado potato beetle is a systematic oligophage in the classification scheme of Otte and Joerne (1977) (Table I), and most of the comparative discussion will also refer to such insects. With respect to *L. decemlineata* the discussion will (1) review its relationship with its original and current host plants; (2) describe and attempt to evaluate recent work on its host-location mechanisms, largely based on the work of Visser and his colleagues (Visser, 1976, 1979b; Visser and Nielsen, 1977; Ma and Visser, 1978; Visser and Ave, 1978; Visser *et al.,* 1979), with consideration given to the evolution of these mechanisms and their possible role in promoting the host-range expansion of the Colorado potato beetle; and (3) as far as possible compare the behavior of the Colorado potato beetle to other oligophagous insects.

**TABLE I.** Various Bases for Grouping Phytophagous Insects into Mono-, Oligo-, and Polyphagous Categories

| Insect Taxa Studied | Monophagous | Oligophagous | Polyphagous | References |
|---|---|---|---|---|
| Broad basis | Feeding restricted to 1 plant species (e.g., members of Homoptera and sawflies) | Feeding usually restricted to 1 plant family (e.g., *Pieris rapae*) | Wide range of host plants, of many families and even orders | Chapman (1971) |
| Moth and butterfly larvae | Normally confined to 1 plant family, or 2 | Feeding on 2–10 plant families | Feeding on 11 or more plant families | Krieger *et al*. (1971) |
| Butterfly larvae | Feeding on only 1 plant genus | Feeding on >1 plant genus, but within 1 order (presumably on >1 family) | Feeding on >1 plant order | Slansky (1976) |
| Grasshoppers | 1. Specific monophagy: feeding restricted to 1 plant species<br>2. Subgeneric monophagy: feeding on several plant species, but within 1 section of a specific genus<br>3. Generic monophagy: feeding on many or all species of 1 plant genus | 1. Systematic oligophagy; feeding on several related genera<br>2. Disjunctive oligophagy: plant genera not related closely<br>3. Sequential or temporary oligophagy: different developmental stages of the insect attack different, unrelated plants | Feeding on different plant orders indiscriminately | Otte and Joern (1977) |

## II. THE COLORADO POTATO BEETLE AND ITS HOST PLANTS

### A. Historical Perspective

The original hosts and the course of geographic spread of *L. decemlineata* are better known than almost any comparable pest mainly because of the work of Tower (1906). There is general agreement that the buffalo bur *Solanum rostratum* was the principal North American host of Colorado potato beetles before the introduction of potatoes (*Solanum tuberosum*). Other *Solanum* species, probably including *S. eleagnifolium,* were occasionally used, but apparently the beetles were entirely restricted to this genus and largely to the subgenus *Leptostemonum* (Hsiao, 1981).

Tower (1906) speculated that *Solanum rostratum* and, consequently, *L. decemlineata,* originated in southern or central Mexico and was spread to the southeastern slope of the Rocky Mountains by Spanish travelers. The closest relatives of *S. rostratum* are all Mexican (Whalen, 1979). However that may be, the beetles had begun to attack potatoes in the western Great Plains by 1858 and began a rapid expansion of their geographic range, reaching the eastern seaboard of the United States by 1873 and covering much of the United States and southern Canada by about 1900 (Tower, 1906). The direction of expansion usually, but not invariably, followed the prevailing winds of the region (Johnson, 1969). The beetle first became established in Europe (France) in 1921. It has subsequently spread north to the Baltic, eastward into the Soviet Union, and southeastward into Turkey (de Wilde and Hsiao, 1981). Throughout this vast range its major host is potato, but in North America several other crops are attacked, as described in the next section.

### B. Current Host Plants

Hsiao (1974, 1978, 1981) has described the natural host range of *L. decemlineata* and related species of *Leptinotarsa;* his results are summarized in Table II. Several other plants, including a few nonsolanaceous species, support feeding and growth, but are not attacked in the field (Hsiao and Fraenkel, 1968b).

Because the host-finding adaptations of the Colorado potato beetle probably evolved largely in response to the properties of the plants on which it originally fed, a brief description of these, especially *Solanum rostratum,* is in order. Buffalo bur is found in open, semiarid grassland, or along roadsides and in waste ground (Whalen, 1979) where vegetation is comparatively sparse and species diversity is low or moderate. It is a prickly annual herb, but individual plants

**TABLE II.** Host Plants (Family Solanaceae) of the Colorado Potato Beetle *Leptinotarsa decemlineata*[a]

| Plant Genus | Species | Suitability[b] |
|---|---|---|
| *Solanum*[c] | *S. tuberosum* | ++++ |
| | *S. dulcamara* | ++++ |
| | *S. rostratum* | ++++ |
| | *S. carolinense* | +++ |
| | *S. elaeagnifolium* | +++ |
| | *S. melongena* | ++++ |
| | *S. sarachoides* | +++ |
| | *S. trifolium* | ++ |
| *Hyoscyamus* | *H. niger* | +++ |
| *Lycopersicon*[d] | *L. esculentum* (certain varieties only) | +++ |
| *Capsicum*[e] | *C. annuum* | + |

[a]Modified after Hsiao (1981).
[b]++++, optimal growth and reproduction; +++, moderate growth and reproduction; ++, slight growth and reproduction; +, restricted feeding by adult beetles only.
[c]Belonging to three subgenera: *Potatoe,* first two species; *Leptostemonum,* next four species; and *Solanum,* last two species.
[d]Feeding and economic damage has reportedly increased in recent years on tomato plants (Schalk and Stoner, 1979). The beetles, however, still prefer stems and fruits over foliage, suggesting incomplete adaptation, presumably to higher concentrations of toxic alkaloid, tomatine, in the foliage.
[e]Damage observed only since 1980 (J. H. Lashomb, personal communication, 1981). Damage confined only to peppers immediately after transplantation, and limited to feeding on stems rather than on foliage. Moreover, the feeding is by adults only; the larvae do not develop on this species. This could indicate a trend toward broadening of the beetle's host range, although the present adaptation is somewhat limited, as for tomato.

commonly reach the stature of small shrubs. It is evidently adapted to opportunistic exploitation of shifting habitats and has become a weed throughout large areas of North America and on other continents (Whalen, 1979; Hsiao, 1981). Thus its local distribution is likely to be patchy, but the patches can be large. Bowers (1975) described her study plots as containing up to 500 densely clustered plants. Plants are present at a site for several months during each summer (Ricketts, 1968; Bowers, 1975).

Most of the other natural food plants are also somewhat "weedy" and occur in open ground, often in groups. For example, *Solanum carolinense* and *S. elaeagnifolium* are perennial weeds of sparsely vegetated areas; they spread vegetatively, again forming fairly large but often widely dispersed patches (M. L. May, unpublished observations, 1981).

As a group, these plants are visually rather divergent as to individual size, color, and leaf shape. Because they are mostly fairly closely related, they presumably share common chemical characteristics, but apparently little or nothing is known of the volatile constituents of their foliage. Hsiao (1981) has suggested a coevolutionary relationship of *L. decemlineata* and its close relatives with *Solanum* of the *Leptostemonum* group.

Notwithstanding this coevolution, the Colorado potato beetle now is highly adapted physiologically and behaviorally to potato. From the speed with which the insect spread once potato became available to it, it must have been preadapted to utilize *S. tuberosum,* although some subsequent adaptation to this host is also likely (e.g., European Colorado potato beetles now have a higher rate of survival to adulthood on *S. tuberosum* than on *S. rostratum*) (de Wilde and Hsiao, 1981). Hsiao (1978) noted, however, that *S. rostratum,* even though it supports growth and reproduction no better than potato, is still distinctly preferred to any other solanaceous plant as an oviposition site by the North American beetles. The beetles also do quite well on two other exotic hosts: bittersweet (*S. dulcamara*) and eggplant (*S. melongena*). Nevertheless, in the state of New Jersey at least, *S. dulcamara* is rarely heavily infested despite its abundance near large Colorado potato beetle populations. Eggplant supports reproducing populations but is most seriously attacked by emigrating adults after potatoes begin to senesce (J. H. Lashomb, personal communication, 1981). Because in the laboratory these plants are nearly equal to potato in acceptability (Hsiao and Fraenkel, 1968b; Hsiao, 1974), the apparent preference for potato in the field may relate either to the other plant species' dispersion or phenology, or to differential attraction of the insects from a distance. Several other plants are attacked, as noted in Table II, but both laboratory and field evidence show that these support feeding and growth less well than those already discussed.

Hsiao (1978) has also demonstrated distinct differences among geographically isolated Colorado potato beetle populations. Four populations from the southwestern United States differed in the spectrum of their responses (including growth, developmental rate, mortality, fecundity, and diapause) to a variety of possible host plants. One population, from Arizona, was particularly distinctive in its ability to utilize *S. elaeagnifolium,* on which it was collected, as well as *S. tuberosum*. For other populations *S. elaeagnifolium* was a distinctly inferior host. Hsiao suggested that the various Colorado potato beetle biotypes may differ in host-location behavior as well as responses to feeding deterrents and toxicants, and that evolution of such differences was likely to occur only when the Colorado potato beetle populations were relatively isolated from each other and from the normally preferred hosts. Distinct host races do not appear to have evolved in Europe (de Wilde and Hsiao, 1981), although one laboratory strain developed more successfully on a greater variety of food plants than any wild population, showing that evolution of these characteristics can be fairly rapid.

### C. Basis of Host-Plant Acceptance

Although the major emphasis in this chapter is in host location at a distance, part of the process of finding suitable plants by adults, and continued acceptance by adults and larvae, clearly depends on contact stimuli. The various important factors for the Colorado potato beetle have been summarized by Hsiao (1969). Once host contact has been made, recognition appears to be potentiated by the presence of chemical sign stimuli that are largely specific to Solanaceae. The nature of these stimuli is uncertain; Chauvin (1952) suggested that phenolic flavonoids were responsible, but Hsiao and Fraenkel (1968a,d) were unable to isolate a specific compound and suggested that a mixture of substances could be involved. The presence of such stimuli is not absolutely essential, however, because even in their absence, after a period of adaptation, feeding could be initiated on nonsolanaceous plants or an artificial diet (Hsiao, 1969). Feeding only occurred in the presence of one or more stimulative nutrient chemicals, including sugars, amino acids, and several others common to most plants (Hsiao and Fraenkel, 1968c). To a large extent, feeding specificity is determined by the presence or absence of repellents, including several alkaloids of some members of Solanaceae, such as demissin, leptine, and tomatine (Schreiber, 1958). A large number of other plants have been shown to have repellent properties (Jermy, 1958, 1961; Hsiao and Fraenkel, 1968b).

The work of Hsiao and Fraenkel (1968a,b,c,d) refers mostly to larvae and that of Jermy (1958, 1961) mainly to adults. Their results are similar, and it is assumed that adults and larvae respond to the same contact cues for feeding. Oviposition by adult females is the point at which the most important host-selection behavior takes place, however, and to some degree this is influenced by different stimuli than feeding. In choice tests, females discriminate between solanaceaeous and nonsolanaceous plants, and among species of Solanaceae. The preference for *Solanum rostratum* has already been mentioned. Some nonsolanaceous plants are somewhat acceptable whereas some Solanaceae are rejected because of the presence of chemical inhibitors (Hsiao and Fraenkel, 1968b; Hsiao, 1969). Interestingly, *Solanum nigrum* (*S. americanum* ?) (Heiser *et al.*, 1979) and *S. luteum* are chosen over *S. tuberosum* as oviposition substrates even though they do not support feeding and larval growth (de Wilde *et al.*, 1960; Hsiao and Fraenkel, 1968b). Oviposition on nonsolanaceous weeds growing in potato fields is not uncommon (M. L. May, unpublished observations, 1981).

### D. Dispersal

The ability of any insect to find suitable hosts depends not only on its sensory capacities and the characteristics of the plant, but also on the insect's mobility and pattern of dispersal. Although it does not appear to be a strong or frequent flier (Johnson, 1969), the Colorado potato beetle clearly is a highly mobile insect capable of wide dispersal when the species as a whole is examined over a time scale of years and on a continental scale in space. This is obviously in part a consequence of its suddenly having access to vast areas of a previously unexploited resource, the potato. Comparatively little is known about the details of movement and searching patterns of individual adults (Johnson, 1969).

In North America and southern Europe, migration flights usually take place before diapause in the fall. Postdiapause spring migration has been commonly recorded in northern Europe, and at this time flight is often inhibited by low temperature or lack of insolation (upon which body temperature is strongly dependent) (May, 1981). Migratory flights are commonly wind assisted; thus major movements are downwind, although local upwind flights during migration as well as at other times are, of course, possible. Movements of up to several hundred meters are often made in spring entirely by walking (Jermy, 1958). Johnson (1969), following Tower (1906), related these dispersal movements to ovarian development, suggesting that migrations occur, weather permitting, toward the end of the preoviposition period. Nevertheless, it is apparent that some mass movements in the fall are direct responses to removal of host plants by the harvesting of potatoes. At this time vast numbers of beetles disperse and become much more common on alternative solanaceous hosts. Many, however, fail to find suitable food at this time and die, whereas others may diapause immediately (J. H. Lashomb, personal communication, 1981).

Even during summer when food is abundant, a small but noticeable fraction of beetles take flight from potato fields beginning near midday. This flight may be keyed by temperature (J. H. Lashomb, personal communication, 1981), but nothing is known of the sex ratio, physiological state, or subsequent flight behavior of these beetles. It is interesting to speculate whether such flights may not have evolved in the context of location of more widely dispersed endemic hosts.

## III. HOST-SEEKING BEHAVIOR OF THE COLORADO POTATO BEETLE

### A. Vision

Long-range host location, except by random search, must be mediated at least in part by vision or olfaction or both. Although evidence from the recent literature (e.g., Prokopy and Owens, 1978; Rausher, 1978) and elsewhere in this volume emphasizes the importance of visual orientation in many insects, this aspect of response has been little investigated in the Colorado potato beetle. De

Wilde and Pet (1957) showed that larvae are attracted to green in comparison to red or ochre, but only over short distances. Apparently adult vision has not been examined even to this extent. Casual field observations suggest that adults walking on the ground are attracted to vertical objects (J. H. Lashomb, personal communication, 1981), but this awaits confirmation.

### B. Olfaction

Because oviposition choice by adult females largely determines the larval host, it is to be expected that olfaction is most prominent in adults. Nevertheless, Chin (1950) showed in "screen tests" that larvae can orient to solanaceous leaves, and that this ability is abolished if the antennae are amputated. Because the tests were conducted in still air, this may represent true chemotaxis or may merely be a kinetic response. De Wilde (1958) described regular side-to-side movements of the heads of searching larvae; although he ascribed to this behavior a role in phototaxis, it could also serve in chemotaxis. In these experiments, detection of food plants was possible only from a few centimeters away. It is, of course, unlikely that larvae would often have to locate a host at greater distances. The possibility of chemically mediated anemotaxis, as described later for adults, has not been ruled out.

Olfactory responses to plants have been studied perhaps more thoroughly in adult Colorado potato beetles than in any other insect. As early as 1926, McIndoo used a Y-tube olfactometer to demonstrate attraction to unbruised potato foliage, and Schanz (1953) later showed that the response was not merely one to humidity. In her experiments the attraction to potato was largely abolished by excision of the terminal five antennal segments.

The limitations of typical olfactometers have been outlined by Dethier (1947) and Kennedy (1977b). For example, steep gradients of odorants may exist in an olfactometer, thus permitting chemotactic orientation. Because this can probably operate only over very short distances in walking insects (Kennedy, 1977a,b), its relevance to location of distant hosts is questionable. Odor-conditioned anemotaxis is much more likely to be of practical significance for orientation at a distance in the field (Kennedy, 1977a,b). This mode of orientation has been studied in wind tunnel experiments. De Wilde *et al.* (1969b) were the first to test Colorado potato beetle responses to plant odors in a wind tunnel. Their analysis demonstrated clear positive anemotaxis in walking adult beetles, and this taxis was substantially enhanced by the odor of leaves of several solanaceous plants. Starvation slightly increased the response, whereas removal of the terminal four segments of the antennae abolished oriented response. Nonsolanaceous plant leaves were either neutral or repellent except that celery (*Apium graveolens*) was slightly attractive in some tests, neutral in others. These results are summarized in Table III.

**TABLE III.** Attractivity, Neutrality, and Repellency of Foliage of Some Plant Species to *Leptinotarsa decemlineata*[a]

| Family and Species | References |
|---|---|
| Attractive plants | |
| Umbelliferae | |
| *Apium graveolens* | de Wilde *et al.* (1969b) |
| Tropaeolaceae | |
| *Tropaeolum majus* | Visser and Nielsen (1977) |
| Solanaecae | |
| *Solanum tuberosum* | McIndoo (1926); Chin (1950); Schanz (1953); Grison (1957); Jermy (1958); de Wilde *et al.* (1969b); Visser and Nielsen (1977) |
| *Solanum dulcamara* | Schanz (1953); de Wilde *et al.* (1969b); Visser and Nielsen (1977) |
| *Solanum luteum* | Schanz (1953) |
| *Solanum nigrum* | Schanz (1953); de Wilde *et al.* (1969b); Visser and Nielsen (1977) |
| *Solanum carolinense* | McIndoo (1926) |
| *Solanum demissum* | Bongers (1970) |
| *Nicotiana tabacum* | Visser and Nielsen (1977) |
| *Lycopersicon esculentum* | McIndoo (1926); Schanz (1953); de Wilde *et al.* (1969b); Visser and Nielsen (1977) |
| *Capsicum annuum* | Visser and Nielsen (1977) |
| *Petunia hybrida* | Chin (1950); Grison (1957); Jermy (1958); Bongers (1970); Visser and Nielsen (1977) |
| *Hyoscyamus niger* | McIndoo (1926) |
| *Datura stramonium* | McIndoo (1926) |
| Neutral plants | |
| Corylaceae | |
| *Alnus incana* | de Wilde *et al.* (1969b) |
| Papillionaceae | |
| *Medicago sativa* | Grison (1957) |
| *Pisum sativum* | Chin (1950) |
| *Phaseolus vulgaris* | Visser and Nielsen (1977) |
| Cruciferae | |
| *Brassica* spp. | Chin (1950) |
| *Brassica oleracea* | Visser and Nielsen (1977) |
| *Brassica pekinensis* | Visser and Nielsen (1977) |
| *Rhaphanus sativus* | Visser and Nielsen (1977) |
| Compositae | |
| *Doronicum pardalianches* | Bongers (1970) |
| Repellent plants | |
| Compositae | |
| *Taraxacum officinale* | de Wilde *et al.* (1969b) |
| Graminae | |
| *Poa annua* | de Wilde *et al.* (1969b) |

[a]Modified from Visser and Nielsen (1977).

Visser and his co-workers extended these studies using a carefully designed wind tunnel that produced laminar airflow over a "walking plate" on which adult beetles were placed and where the speed and direction of walking could be accurately assessed (Visser, 1976). Temperature, humidity, and distribution of light were measured and controlled. Visser (1976) confirmed that the beetles exhibit significant positive anemotaxis in clean (charcoal-filtered) air. When intact potato plants were positioned upwind in the tunnel in such a way that odors were uniform across the tunnel cross section, a far higher proportion of beetles collected at the upwind end of the plate. Because no odor gradients existed within the tunnel, this response must have been an odor-conditioned anemotaxis. The response was further enhanced somewhat by prolonged starvation and by use of older (4- to 8-week-old) plants, which produce higher concentrations of attractive volatiles (Visser, 1979b). Fully fed beetles are not attracted (J. H. Visser, personal communication, 1981). Visser and Nielsen (1977) showed that a variety of other solanaceous plants markedly enhanced anemotaxis in unfed, newly emerged Colorado potato beetles (Fig. 1), as well as inducing distinct orthokinesis; beetles reached the edge of the plate more rapidly when the plant odors were present. Their results are included in Table III. Several solanaceaous plants that are unacceptable or marginal for feeding (e.g., *Nicotiana tabacum, Capsicum annum,* and *Petunia hybrida*) are nevertheless highly attractive. Most

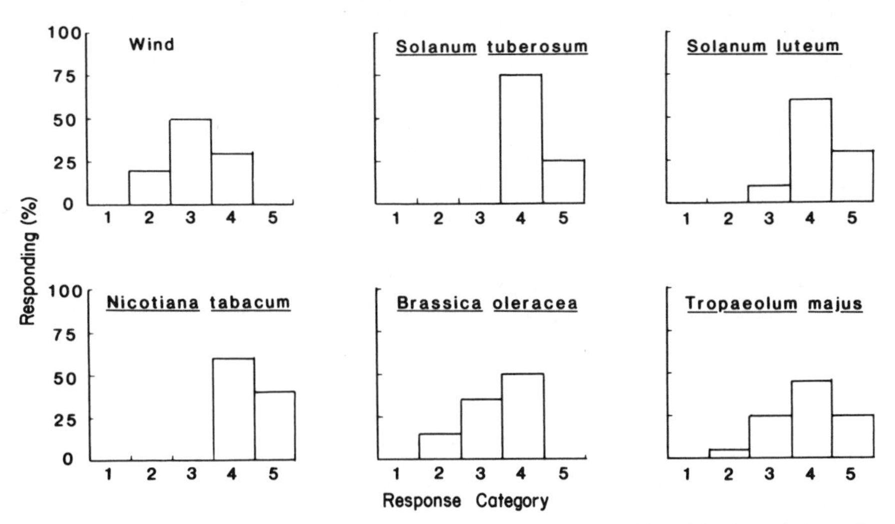


**Fig. 1.** Responses of unfed, newly emerged female Colorado potato beetles walking in a wind tunnel to wind and to the scents of selected solanaceous and nonsolanaceous plants. Categories of response: (1) moved directly to downwind side of walking plate; (2) moved indirectly to downwind side; (3) moved to lateral edges; (4) moved indirectly to upwind side; (5) moved directly to upwind side. (Data from Table III of Visser and Nielsen, 1977.)

nonsolanaceous plants tested by Visser and Nielsen were not significantly attractive, although for all of these the mean category of response (scale of 1 to 5 from direct downwind to direct upwind) was somewhat higher than for wind alone. One exception was the garden nasturtium *Tropaeolum majus,* which did elicit a significant degree of attraction and orthokinesis. When beetles were fed on potato or on bittersweet for 48 hr, then starved for 24 hr, responses to plants were generally similar to responses of beetles that had never fed as adults, although potato-fed beetles responded to bittersweet less rapidly than to potato.

The nature of the attractants in potato has also received substantial attention. The volatile constituents of *S. tuberosum* foliage were isolated, and the main components were identified as *trans*-2-hexen-1-ol, 1-hexanol, *cis*-3-hexen-1-ol, *trans*-2-hexenal, and linalool; relative concentrations in the tissues were 100:17:7:7:4, respectively (Visser *et al.,* 1979). The main headspace vapors over cut potato leaves are *cis*-3-hexen-1-ol, *cis*-3-hexenylacetate, *trans*-2-hexenal, and *trans*-2-hexen-1-ol in a ratio of 100:59:37:16 (Visser and Ave, 1978), and some unidentified sesquiterpenes (Visser, 1979b).

The main components identified in the essential oil of potato leaves (Visser *et al.,* 1979 and references therein) are widely distributed in plant leaves. These compounds are smelled by humans as a grasslike odor and are termed *green-leaf volatiles* by Visser and Ave (1978). The straight-chain saturated and unsaturated aldehydes and alcohols are formed by oxidative degradation of plant lipids, as shown in Fig. 2. The relative proportion of these end products (mostly alcohols and aldehydes) varies among different plant species within the same genus, and the proportions are modified seasonally within the same species owing to plant aging and injury (Buttery *et al.,* 1969; Kazeniac and Hall, 1970; Sayo and

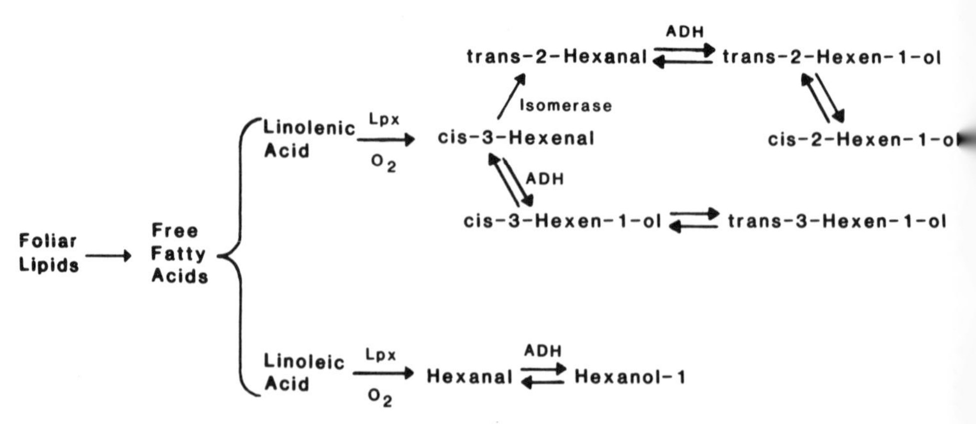


**Fig. 2.** Proposed biosynthesis of green-leaf volatile complex of leaf aldehydes and alcohols. Lpx; lipoxygenase; ADH; alcohol dehydrogenase. (Modified from Visser *et al.,* 1979, where sources of original data appear.)

Takeo, 1975; Hatanaka *et al.*, 1976; Visser, 1979b). Visser *et al.* (1979) presented an extensive compilation of plant families, leaves of which contain the same volatiles that predominate in potato foliage. At least 11 families contain all four of the major alcohols and aldehydes. Many of these (for example, Rutaceae, Umbelliferae, and Labiatae) also contain other strong odorants not detected in potato, however, and in the absence of quantitative data to the contrary it is reasonable to assume that the proportions of the green-leaf volatiles are, on the whole, more similar within the Solanaceae than between families.

The electroantennogram (EAG) responses of Colorado potato beetles to these and other generally occurring plant volatiles (a total of 53 compounds) also were assessed (Visser, 1979b). Colorado potato beetles' antennae were most sensitive to a group of closely related compounds—namely, the general green-leaf volatiles *trans*-2-hexen-1-ol, *cis*-3-hexen-1-ol, 1-hexanol, *trans*-2-hexenal, and *cis*-3-hexenylacetate, and to isomers such as *trans*-3-hexen-1-ol and *cis*-2-hexen-1-ol (Table IV). Methylsalicylate also evoked definite responses, though

**TABLE IV.** EAG Responses of Potato Beetle to Test Compounds ($10^{-3}$ *M* in Paraffin Oil v/v) as a Percentage of the Response to *cis*-3-hexen-1-ol[a]

| Compound | Response | Compound | Response |
|---|---|---|---|
| Propionic acid | 0 | Butylacetate | 1 |
| Butanoic acid | 0 | Benzylacetate | 7 |
| Pentanoic acid | 0 | Benzaldehyde | 5 |
| Hexanoic acid | 0 | Salicylaldehyde | 6 |
| Heptanoic acid | 0 | Methylsalicylate | 22 |
| Linoleic acid | 0 | Eugenol | 9 |
| Hexanal | 21 | $\Delta^3$-Carene | 0 |
| 1-Hexanol | 94 | (+)-Limonene | 0 |
| 2-Hexanol | 17 | α-Pinene | 0 |
| 3-Hexanol | 33 | β-Ionone | 0 |
| *trans*-2-Hexenal | 39 | Geraniol | 0 |
| *trans*-2-Hexen-1-ol | 146 | Nerol | 0 |
| *cis*-2-Hexen-1-ol | 75 | Linalool | 18 |
| *cis*-3-Hexen-1-ol | 100 | Citronellol | 4 |
| *trans*-3-Hexen-1-ol | 101 | 2-(Methylthio)ethanol | 2 |
| *cis*-3-Hexenylacetate | 69 | 3-(Methylthio)propanal | 4 |
| 2-Methylbutanol-1 | 0 | 2-Pentanone | 6 |
| 3-Methylbutanol-1 | 1 | 6-Methyl-5-Hepten-2-one | 7 |
| 1-Penten-3-ol | 6 | 2-Hendecanone | 1 |
| 5-Hexen-1-ol | 37 | Pyridine | 6 |
| 1-Octen-3-ol | 57 | | |

[a]Data from Visser (1979b).

small compared to *trans*-2-hexen-1-ol. Sensitivity to *trans*-2-hexen-1-ol (concentration of $1 \times 10^{-6} M$ v/v in paraffin oil; equivalent to $1.2 \times 10^8$ molecules/$cm^3$ of air) was greatest, followed by *trans*-3-hexen-1-ol (concentration, $1 \times 10^{-6} M$ v/v; equivalent to $2.3 \times 10^8$ molecules/$cm^3$ of air). In comparison, at 760 mm Hg and 20°C, 1 $cm^3$ of air contains about $10^{19}$ molecules. Thus sensitivity to potato leaf volatiles was clearly established. The beetles respond strongly to all of the major volatiles identified in the essential oil of the potato leaves, with the exception of the relatively small reaction to linalool. At the EAG level, however, specificity to particular plants is not evident, because leaf vapors of a variety of solanaceous and nonsolanaceous plants elicited distinct and roughly equal responses, presumably in part because all emitted the general green-leaf volatiles. The nonsolanaceous plants included some (e.g., *Brassica oleracea*) that were previously shown to be unattractive to intact beetles.

Ma and Visser (1978) studied olfactory coding at the single-unit level. Activity in olfactory neurons associated with sensilla basiconica was recorded extracellularly during serial exposure to as many as 13 volatile compounds delivered in air. The test compounds included the six-carbon, straight-chain alcohols and aldehydes found in potato foliage, as well as some additional isomers, plus linalool, methylsalicylate, and α-pinene. At least five different types of receptors were identified by cluster analysis of their response spectra. Four of these were primarily sensitive to various combinations of the green-leaf volatile complex and the fifth primarily to methylsalicylate. In some cases activity could be recorded simultaneously in two neurons of different types in the same sensillum. By comparing ratios of activity in two such units when exposed to actual plant odors, Ma and Visser showed considerable discrimination even by single pairs of receptors. For example, one pair could potentially discriminate plants of the genus *Solanum* from tomato and several nonsolanaceous species. Thus, although the EAG as a whole was not plant specific, the individual receptors showed the potential for making rather fine distinctions.

Finally, Visser and Ave (1978) investigated the effect on attraction of Colorado potato beetles to intact potato plants of adding additional amounts of potato leaf volatiles to the airstream. First they found that in the absence of plants the beetles were neither attracted nor repelled by the individual volatile compounds alone. In several cases, however, addition of excess volatiles to an airstream passing over a potato plant strongly reduced the odor-conditioned anemotaxis of the insects. This was true when *trans*-3-hexen-1-ol, *cis*-2-hexen-1-ol, *trans*-2-hexen-1-ol, or *trans*-2-hexenal were added; these chemicals had less effect on the degree of chemo-orthokinesis shown by the beetles. Only the latter two compounds have actually been detected in potato essential oil or headspace of cut potato leaves. Visser and Ave (1978) concluded that the effect of these added volatiles was to mask the naturally attractive odor of potato plants, probably by disrupting the "essence," or precise ratio of volatiles, necessary to enhance anemotaxis.

If, indeed, the potato beetle depends on the specific volatile essence of the potato plant and other solanaceous plants for host location, then it must be able (1) to do so in spite of the changes in volatiles released as a result of seasonal changes of the host plant itself, and (2) to discriminate the overlap of similar green-leaf volatiles given off by many other plants in the field. At present the following possibilities exist: (1) that the necessary components of the essence include all of, or a particular fraction of, the volatiles of the essential oil of potato isolated by Visser *et al.* (1979), and that specificity is conferred by a precise ratio or small range of ratios of these components that is characteristic of Solanaceae; or (2) that in addition to those volatiles there may be one or more components, evidently present in small quantities, that are specific to Solanaceae and that release or potentiate the anemotactic response to the preferred host.

The first of these possibilities is essentially the hypothesis proposed by Visser and Ave (1978) and is a reasonable explanation of their results to date. This supposes a situation strikingly like that found in some pheromone systems (Birch, 1974). The hypothesis is subject, however, to the difficulties noted in the preceding discussion, and it would be of considerable interest to know how narrowly specific such an orientation system could be. In this regard, quantitative information on the relative composition of volatiles emitted by intact plants of different families would be of great interest. Murray *et al.* (1972) isolated from *Solanum campylacanthum* a volatile mixture qualitatively similar to that of potato except for the presence of *cis*-2-hexen-1-ol. Wallbank and Wheatley (1976) found that among crucifers, a plant family whose odor generally is believed to be typified by isothiocyanates, the vapors emitted by disrupted leaf tissues differed rather widely among species, and only in *Brassica oleracea* was allylisothiocyanate prominent. In this case a family-specific ratio of components appears unlikely, and in view of the known attraction of allylisothiocyanate for many crucifer-/feeding insects, the possibility cannot be discounted that the insects are more sensitive to specific rare components than are human analytical methods.

The apparent attractiveness of the nonsolanaceous *Tropaeolum majus* and *Apium graveolens* also raises interesting questions, especially because the essential oils of *Apium* are dominated by terpenoids (de Wilde *et al.*, 1969b), although short-chain alcohols and aldehydes are undoubtedly present (Visser *et al.*, 1979). If attraction to these plants is based on the same mix of volatiles as potato, then these plants must have converged on Solanaceae in this respect, and the attractiveness of the mixture must occur even in the probable presence of other strong odors. The latter is not too unlikely, as the EAG studies of Visser (1979b) showed little sensitivity to terpenes.

Clearly, however, a direct test of the "potato essence hypothesis" is desirable. This might be accomplished by careful identification and quantification of volatile components released by the foliage of intact potato, and of other Solanaceae as well. Synthetic equivalents of the green-leaf volatiles could then be blended to produce essences identical to that of potato or of other plant odors, and the ratios of components manipulated at will, using equipment and background data described by Visser and Ave (1978) and Ma and Visser (1978). The effects of well-defined stimuli would then be testable and could be compared to the attractive effect of intact potato. If mixtures as close as possible to the essence of potato exhibited equal attraction for Colorado potato beetles, and if changes in the ratios of compounds impaired the attractiveness, then the hypothesis would be proven. Otherwise, the possibility that some other specific compound, not as yet detected, might contribute substantially to the long-range orientation of the beetles should be very carefully considered. In that case a comparison of compounds common to *Solanum* and *T. majus* might be of value.

### C. Evolution of Olfactory Orientation

It appears likely, then, that movements of Colorado potato beetles in the field are potentiated by the scent of host plants and guided by the odor-enhanced anemotaxis primarily to solanaceous plants, with little if any discrimination among the species tested, and possibly to a few attractive nonsolanaceous species (e.g., *T. majus* and *A. graveolens*). More specific host selection may result from chemotaxis at close range, as well as the action of feeding incitants, stimulants, and deterrents as described before. This section will briefly recapitulate some possible factors influencing the evolution of these host-finding mechanisms and their possible role in the development of the beetles' status as pests. We assume here that the olfactory orientation described for walking beetles operates in the same way for flying beetles.

First, several characteristics of the beetles and their original hosts must have selected for some means of orienting toward food plants from a distance. The number of acceptable hosts was small and their distribution probably rather unpredictable in space. Because *S. rostratum* is a weedy annual, postdiapausing beetles might easily find themselves far from food plants. Flight costs are unknown for *Leptinotarsa,* but many beetles have high wing loading and high wingbeat frequency, and thus are likely to experience high costs of flight. Certainly Colorado potato beetles appear incapable of rapid flight, and time could be at a premium during aerial host search. These facts all should favor evolution of some means of increasing host-location efficiency. This could be especially true early in the season when plants are small; although little is known of synchronization of Colorado potato beetles with their natural hosts, they often begin to emerge before potatoes in New Jersey (J. H. Lashomb, personal communication, 1981).

Visser (1976) pointed out that turbulence, especially that caused by obstructions like vegetation, may impair correct orientation in the field, and reduction of wind velocity in the boundary layer near the ground, again enhanced by vegetation, may reduce anemotaxis. Thus, even though walking beetles sometimes climb above the boundary layer on small hills and projections (Visser, 1976), it is likely that orientation by them often is possible only over distances of a meter or less, as observed by Jermy (1958). In contrast, much longer range orientation is likely in flight. Also, the open habitats of *S. rostratum* and the native *Solanum* on which the beetles feed around may tend to facilitate olfactory orientation. The fact that *S. rostratum* may often occur in large patches must also have eased the problems of olfaction by providing a wide and intense odor plume that would not be completely disrupted by small-scale turbulence. Under these circumstances, olfactory orientation could be employed to its greatest advantage.

Selective forces favoring the use of a specific mixture of generally occurring chemicals are harder to pinpoint. The most obvious possibility is that Solanaceae do not produce any single volatile compound that is not also produced by many other plants. The analysis of Visser *et al.* (1979) suggests that this could be so. The specificity, then, could only be introduced by differences in the quantities of odorants and thus, in practice, the ratios among components. Such a system could have certain other advantages in terms of information content, as Wright (1964) pointed out for pheromones. For example, variation in ratios of volatile odorants could convey more detailed information on age and physiological conditions of plants than the presence or absence of a single constituent. Whether or not such detailed information would be useful to an insect, especially at the stage of long-distance location, is questionable and may vary from case to case. The effectiveness of a system of host location depending on discrimination of relative quantities of common chemicals would be enhanced in a community of low diversity because of the reduction in background "noise" from emissions of these chemicals by nonhost plants. For the same reason, such a host-finding mechanism could remain relatively effective in agricultural systems. It is probably true that increased community diversity complicates the host-finding task of any oligophage, but the point here is that diversity would be particularly detrimental in cases where quantitative comparisons of common characters are required.

Responses to host essences are likely to have some inherent flexibility, because to respond successfully the beetle must respond to variability from plant to plant and changes over time in the composition of odor cues. This will preclude extreme specialization at the early stages of host location, and indeed Visser and Nielsen (1977) showed that the range of attractive plants is broader than the range of plants that are eventually accepted for feeding. This property is at least permissive of host-range expansion because it reduces one of the behavioral barriers to acceptance of certain newly available hosts (the same may be true, of course, of visual orientation and of attraction to single components if the latter are of somewhat general occurrence). The Colorado potato beetle is well known for its opportunistic expansion of the range of acceptable hosts (see Table II).

This has been possible in part because of the beetle's preadaptation to find a wide variety of solanaceous plants, coupled with the fact that some introduced *Solanum* spp., especially *S. tuberosum, S. melongena,* and *S. dulcamara,* having evolved outside the original range of *L. decemlineata,* lack toxicants and short-range repellents to the beetle. The general attraction to Solanaceae also contributes to its role as a pest of tomatoes and peppers, plants to which it is physiologically rather poorly adapted but that it sometimes attacks.

## IV. COMPARISON WITH HOST LOCATION IN OTHER OLIGOPHAGOUS INSECTS

In most phytophagous insects, host-selection behavior is a catenary process (Schoonhoven, 1968), requiring attraction or acceptance at several stages and often involving several sensory modalities for initiation and maintenance of feeding. Dual discrimination of secondary plant substances (and plant-based visual cues) and of nutrients often is crucial for the establishment of an insect on its host plant (Beck, 1965). Thus much of the process of identification takes place in direct or near-direct contact with the plant, and many insects may lack means of detecting their hosts at distances of more than a few centimeters (Thorsteinson, 1960).

Nevertheless, numerous other insects are attracted by, or at least arrested by, volatile components of plants on which they feed. Many of the data on larvae were summarized by Jones and Coaker (1978). Phytophagous and saprophagous larvae often make oriented (chemotactic?) movements over short distances. In the case of mono- or oligophagous phytophagous species that feed above ground (i.e., generally on leaves), attraction is usually to secondary plant metabolites. In saprophages, below-ground feeders, and at least some above-ground polyphagous species, primary metabolites such as $CO_2$ play a larger role.

Some of the data for adult insect herbivores are described next. In many cases the precise nature of the attractant is unknown, and even more often the nature of the insects' orientation has not been studied. Given the variation in test procedures and overall paucity of data, it is difficult to generalize about host-location mechanisms. In most studies that have investigated the chemical nature of attractants, certain specific key components have proved to be attractive individually (e.g., Feeny *et al.*, 1970; Matsumoto, 1970; Hawkes and Coaker, 1979). Notably, even the polyphagous vegetable weevil *Listroderes obliquus* apparently is attracted specifically to mustard oils, though only about 15% of this weevil's known hosts are crucifers (Matsumoto, 1970). Presumably other volatiles characteristic of certain other portions of its host range are also attractive. The apparent importance of such key compounds may be in part a result of much work having been concentrated on insects that feed upon chemically distinctive plant groups, like crucifers, that have strong specific odorants. Several studies suggest that attraction might be enhanced by more than one component of the scent, although single components are also attractive alone. For instance, the cotton boll weevil *Anthonomus grandis* is attracted to several of the headspace volatiles over cotton plants (Hedin *et al.*, 1976). Larvae of *Bombyx mori* orient toward several of the components of mulberry leaves (Wanatabe, 1958; Hamamura, 1970). Finch (1978) found that female cabbage root flies (*Delia brassicae*) oviposit more frequently on crucifers containing several volatile glucosinolate derivatives than on those with few or none. Leafhoppers (*Empoasca devestans*) and larvae of the beetle *Lema trilineata* are attracted to several, but not all, subfractions of extracts of their host plants (Kogan and Goeden, 1971; Saxena and Saxena, 1974); this could also, of course, be the result of partitioning of a single active principle among different fractions.

Only in the Colorado potato beetle is there direct evidence that a mixture of volatiles might be required as such for attraction. In a few other insects, indirect evidence points to such a possibility. Saxena and Goyal (1978) found that *Papilio demoleus* adults respond to the odor of citrus leaves, but not to several of the major components of citrus essential oil when presented individually; the components were not tested exhaustively, however. Also, three species of ermine moths (*Yponomeuta*) from different hosts displayed distinct, species-specific EAG-response spectra to 18 different volatile plant chemicals (van der Pers, 1978), including several of the green-leaf volatiles (*sensu* Visser and Ave, 1978). Behavioral responses to the individual compounds or to mixtures were not studied. Adult carrot flies (*Psila rosae*) likewise show strong EAG responses to the green-leaf volatiles, but also respond equally to compounds more specifically characteristic of Umbelliferae (Guerin and Visser, 1980). In addition, the mite *Tetranychus urticae* orients toward artificial mixtures that mimic the composition of the essential oil of strawberry foliage. Attractiveness of the mixture was reduced by reduction or excess of some components (Rodriguez *et al.*, 1976).

Only a few clear cases, other than the Colorado potato beetle, of odor-conditioned anemotaxes toward plant odors are recorded, although, of course, such responses to pheromones are commonplace (e.g., Kennedy and Marsh, 1974). These examples include hoppers of the polyphagous *Schistocerca gregaria* (Kennedy and Moorehouse, 1969) and flying adults of the oligophagous fly *Delia brassicae* (Hawkes and Coaker, 1979) and of the butterfly *Agraulis vanillae* (Copp and Davenport, 1978). Kennedy (1977a,b) made the point, however, that the design of apparatus is critical in revealing and discriminating among various modes of orientation, and no records other then those already listed are available of insects having been tested in situations in which they could move freely in a nearly uniform airflow and odor distribution. Thus odor-conditioned anemotaxis might be widespread among insects that detect food plants by olfaction. This belief is supported by less direct evidence. Field experiments using visually concealed cotton plants showed that boll weevils nevertheless flew upwind toward the plants (Mistric and Mitchell, 1966). The trapping of large numbers of flea beetles (*Phyllotreta*) in small traps baited with plant scents (Feeny *et al.*, 1970) certainly suggests that the influence of the trap extends a long distance, so odor-conditioned anemotaxis may be suspected. Kennedy and Moorehouse (1969) also showed clearly that stimulation of kinetic responses, although in principle entirely independent of olfaction, greatly enhanced the anemotactic response. Visser and Ave (1978) likewise took note of these two separate responses. Flying insects, however, probably are always in a state of activity that could permit subsequent taxes toward food.

Vision plays an important role in host finding of many insects of varying food habits. In many cases it operates along with, perhaps largely as an adjunct to, olfaction (Coaker and Finch, 1973; Saxena and Goyal, 1978). Many insects, for instance, show generalized attraction to green or yellow colors. Others are attracted to vertical objects (Mulkern, 1969) and thus tend to move onto plants. In other cases, however, visual discrimination can be the major mode of host finding and may be quite specific. *Dacus oleae,* the olive fruit fly, is apparently monophagous, but does not rely on olfaction to locate olive trees; rather, these are identified by the distinctive color of their foliage and the fruits later discovered by color and shape (Prokopy and Haniotakis, 1975, 1976). The aphid *Hyalopterus pruni* also identifies its secondary host *Phragmites* largely by its precise hue. Equally specific examples of form discrimination are discussed in Papaj and Rausher, Chapter 3.

The selective pressures leading to one form or another of host finding are evidently complex. In general, the evolutionary choice between random search, with identification of food plants on contact or at very short range, and long-range orientation—as well as the evolution of specificity of orientation—will depend on some balance between the cost in time and energy of finding the host and the difficulty of detecting or interpreting information from the host plant. Monophagous and oligophagous insects searching for oviposition and/or feeding sites are less likely to encounter suitable plants than their polyphagous counterparts. Therefore, for specialized feeders there should be more intense selection for the ability to detect and locate acceptable hosts at a distance. This selective pressure will be roughly proportional to the apparency (Feeny, 1976) of food plants and thus should be greatest for strict monophages. The logic of this prediction seems strong, but evidence is conflicting. Many of the species known or believed to use long-distance cues are oligophagous, but in aphids, even specialized species, only very general visual cues are used, with selectivity depending on contact with the plant (Kennedy, 1973). In contrast, highly polyphagous species like *Schistocerca* and *Listroderes* use olfactory cues that may operate over a considerable distance.

It has also been suggested here that high cost of locomotion may favor efficient host location at a distance. One factor in this cost is the minimum cost of transport, that is metabolic cost of locomotion/distance covered. Generally this cost is higher for walking than for flying, and for small species than for large (e.g., see Schmidt–Nielsen, 1972). The effect of walking versus flying is complicated by different problems of plant detection. For insects that usually find their hosts by flying, however, large species may be more likely to rely on random search than insects of medium size, because the former can cover more ground in less time and at less cost. Very small insects like aphids, however, may be blown about so readily by wind that their search costs are also low, and oriented flight is often relatively difficult or impossible (Kennedy, 1973). Also, as Hawkes *et al.* (1978) pointed out, the high reproductive capacity of aphids implies that the area searched by the progeny of a single aphid is large even though each individual may move short distances. Hawkes *et al.* (1978) also gave an example of a series of *Brassica* pests that fit the prediction. The aphid *Brevicoryne brassicae* and the butterflies of *Pieris* spp. (which are large and also use gliding flight) respond at a distance only to very general cues, so search is more nearly random than for the medium-sized *Delia brassicae,* which orients by specific olfactory cues. Again the evidence on the whole is ambiguous, however. *Manduca sexta,* a large sphinx moth, appears to respond at a distance to almost any tall plant when searching for oviposition sites but accepts or rejects the site on the basis of odor while still in flight within centimeters of the plant (Yamamoto *et al.*, 1969). The smaller moth *Cidaria albulata* (Douwes, 1968), and the still smaller beetles *Sitona cylindricollis* and *Costelytra zealandica* (Hans and Thorsteinson, 1961; Osborne and Hoyt, 1970), apparently are guided to hosts only by kinetic reactions that operate over distances no greater than about 1 m. *Schistocerca* hoppers, which do respond anemotactically (and thus presumably over longer distances), are large and highly mobile as adults; but they must and do walk as hoppers, and then have higher costs of transport.

Some of the problems involved in receiving and interpreting information from a host plant have been indicated. The perceived information should identify suitable hosts and distinguish them from most nonhost plants. Discrimination need not, indeed rarely will, be perfect at this stage, because final identification will take place at the plant itself. Still, the selective advantage of a particular behavior must increase the more it reduces time and energy lost inspecting nonhosts, as long as it does not fail to detect too many suitable hosts. Thus if one or few traits, either olfactory or visual, are to be used individually to identify hosts, they must be sufficiently characteristic of those hosts to allow adequate discrimination from nonhosts. Such key odor characters clearly exist in many plants, but it is not certain that they do in all or even most. Because many insects probably have evolved the ability to locate host plants using key odors, there may well be pressure on the plants to reduce the latter and emit less distinctive odors.

The ease with which a plant odor is disseminated also affects its utility to insects. Structural complexity and species richness increase both the difficulties and the rewards of efficient long-distance host location (Stanton, 1979; Chapter 4, this volume), because in species-rich plant communities, each species must be relatively rare and generally less apparent against a confusing background. Rowell (1978) suggested that this two-pronged dilemma should select for broadening of feeding niche, yet he found that tropical acridids living in very diverse communities are highly specialized. Most work on host location has focused on insect species from comparatively simple communities, doubtless because these are most likely to become agricultural pests. The role of community structure and other factors (e.g., light intensity, wind intensity, and wind steadiness) that may affect the integrity and distinctiveness of the information channels insects use for orientation deserve additional study.

## V. SUMMARY

Selection pressure for oriented long-range host location by phytophagous insects should increase with decreasing breadth of feeding niche, among other factors. Thus oligophagous insects may face different problems, at least in degree, from those of monophagous or polyphagous species. Here we review host finding in the Colorado potato beetle *Leptinotarsa decemlineata,* and compare it with other oligophagous insects.

The Colorado potato beetle originally fed mostly on *Solanum rostratum* in the southwestern United States and Mexico. Within historical times it has expanded its host range to include several solanaceous crops, notably potato, and its geographic range over much of North America and Europe. Host-finding mechanisms presumably evolved largely before this range expansion, so some characteristics of the beetle's hosts and its dispersal behavior are briefly reviewed. Final host-plant selection depends strongly on contact stimuli from the plants, probably including unidentified compounds characteristic of Solanaceae that act as biting stimulants, phagostimulative nutrients widespread in plants generally, and the presence or absence of a wide array of repellent or rejectant plant allelochemicals.

Visser and his colleagues in the Netherlands have extensively studied beetles' detection of and orientation toward distant hosts. The insects move slightly upwind in laminar airflow in a wind tunnel. This tendency is greatly enhanced by the odor of intact potato or other solanaceous plants. With a few exceptions, nonsolanaceous plants show little or no attractivity. The major volatile components of potato foliage are mostly six-carbon alcohols and aldehydes common to many plants. Colorado potato beetles show strong EAG responses to several of these compounds but are neither attracted nor repelled by the compounds individually. Nevertheless, some of these compounds, if added in excess along with natural potato odor, depress the odor-conditioned anemotaxis excited by potato. These facts led to the proposal that the attraction to potato was actually due to an essence consisting of several volatile compounds in a particular ratio characteristic of Solanaceae. This hypothesis has not been tested directly. Whether or not adequate specificity can be encoded in this way remains an open question.

For most other insects in which chemical attractants have been identified, these consist of specific key substances to which the insect responds. In a few cases mixtures of compounds may enhance attractivity. Most work, however, concerns insects that feed upon plants having highly distinctive odorants; very little direct evidence bears on the role of mixed essences. Few experiments have been designed to test the existence of plant odor-conditioned anemotaxis, although studies on pheromones suggest that this is the olfactory mechanism most likely to be effective at long distance.

Clearly much experimental work remains to be done. We predict that mechanisms of host location depend on the breadth of the feeding niche, the mobility of the insect, the abundance and dispersion of host plants, and the chemical distinctiveness of hosts relative to other plants in the habitat. Present data are inadequate to test any of these predictions.

## Acknowledgments

Our sincere thanks to Dr. J. H. Lashomb (Department of Entomology and Economic Zoology, Cook College, Rutgers University) for sharing with us his observations of the Colorado potato beetle in the field, and to Dr. J. H. Visser (Department of Entomology, Agricultural University, Wageningen) for kindly clarifying for us several aspects of his work.

## REFERENCES

Beck, S. D. (1965). Resistance of plants to insects. *Annu. Rev. Entomol.* **10,** 207–232.

Birch, M. C. (1974). Pheromones. *Front. Biol.* **32,** 1–7.

Bongers, W. (1970). Aspects of host–plant relationship of the Colorado beetle. *Meded. Landbouwhogesch. Wageningen,* No. 70–10.

Bowers, K. A. W. (1975). The pollination ecology of Solanum rostratum (Solanaceae). *Am. J. Bot.* **62,** 633–638.

Buttery, R. G., Seifert, R. M., Guadagni, D. G., and Ling, L. C. (1969). Characterization of some volatile constituents of bell peppers. *J. Agric. Food Chem.* **17,** 1322–1327.

Chapman, R. F. (1971). "The Insects: Structure and Function," 2nd ed. Elsevier, New York.

Chauvin, R. (1952). Nouvelles recherches sur les substances qui attirent le doryphore (*Leptinotarsa decemlineata* Say) vers la pomme de terre. *Ann. Epiphyt.* **3,** 303–308.

Chin, C. T. (1950). Studies on the physiological relation between the larvae of *Leptinotarsa decemlineata* (Coleoptera: Chrysomelidae) and some solanaceous plants. *Tijdschr. Plantenziekten* **56,** 1–88.

Coaker, T. H., and Finch, S. (1973). The association of the cabbage rootfly with its food and host–plants. *Symp. R. Entomol. Soc. London* **6,** 119–128.

Copp, N. H., and Davenport, D. (1978). *Agraulis* and *Passiflora*. II. Behavior and sensory modalities. *Biol. Bull.* **155,** 113–124.

Dethier, V. G. (1947). "Chemical Insect Attractants." Blakiston, Philadelphia.

Dethier, V. G. (1978). Studies on insect/plant relations—past and future. *Entomol. Exp. Appl.* **24,** 759–766.

de Wilde, J. (1958). Host plant selection in the Colorado beetle larva (*Leptinotarsa decemlineata* Say). *Entomol. Exp. Appl.* **1,** 14–22.

de Wilde, J., and Hsiao, T. (1981). Geographic diversity of the Colorado potato beetle and its infestation in Eurasia. *In* "Advances in Potato Pest Management" (J. H. Lashomb, and R. A. Casagrade, eds.), pp. 47–68. Hutchinson & Ross, Stroudsburg, Pennsylvania.

de Wilde, J., and Pet, J. (1957). Optical components in the orientation of the Colorado beetle larva (*Leptinotarsa decemlineata* Say). *Acta Physiol. Pharmacol. Neerl.* **6,** 715–726.

de Wilde, J., Bongers, W., and Schooneveld, H. (1969a). Effects of host plant age on phytophagous insects. *Entomol. Exp. Appl.* **12,** 714–720.

de Wilde, J., Hille Ris Lambers–Suverkropp, K., and van Tol, A. (1969b). Responses to air flow and air-borne plant odour in the Colorado beetle. *Neth. J. Plant Pathol.* **75,** 53–57.

de Wilde, J., Sloof, R., and Bongers, W. (1960). A comparative study of feeding and oviposition preferences in the Colorado beetle (*Leptinotarsa decemlineata* Say) *Meded. Landbouwhogesch. Opzoekingsstn. Staat Gent* **25,** 1340–1346.

Douwes, P. (1968). Host selection and host finding in the egg-laying female *Cidaria albulata* L. (Lep. Geometridae). *Opusc. Entomol.* **33,** 233–279.

Feeny, P. (1976). Plant apparency and chemical defense. *Recent Adv. Phytochem.* **10,** 1–40.

Feeny, P., Paauwe, K. L., and Demong, N. J. (1970). Flea beetles and mustard oils: Host plant specificity of *Phyllotreta cruciferae* and *P. striolata* adults (Coleoptera: Chrysomelidae). *Ann. Entomol. Soc. Am.* **63,** 832–841.

Finch, S. (1978). Volatile plant chemicals and their effect on host finding by the cabbage root fly (*Delia brassicae*). *Entomol. Exp. Appl.* **24,** 350–359.

Fox, L. R., and Morrow, P. A. (1981). Specialization: Species property or local phenomenon? *Science* **211,** 887–893.

Grison, P. (1957). Les facteurs alimentaires de la fécondité chez le doryphore (*Leptinotarsa decemlineata* Say) (Coleptera: Chrysomelidae). *Ann. Epiphyt.* **8,** 304–381.

Guerin, P. M., and Visser, J. H. (1980). Electroantennogram responses of the carrot fly, *Psila rosae,* to volatile plant components. *Physiol. Entomol.* **5,** 111–119.

Hamamura, Y. (1970). The substances that control the feeding behavior and the growth of the silkworm *Bombyx mori*. *In* "Control of Insect Behavior by Natural Products" (D. L. Wood, R. M. Silverstein, and M. Nakajima, eds.), pp. 55–80. Academic Press, New York.

Hans, H., and Thorsteinson, A. J. (1961). The influence of physical factors and host plant odour on the induction and termination of dispersal flights in *Sitona cylindricollis* Fabr. *Entomol. Exp. Appl.* **4,** 165–177.

Hatanaka, A., Kajiwara, T., and Sekiya, J. (1976). Seasonal variations in *trans*-2-hexenal and linolenic acid in homogenates of *Thea sinensis* leaves. *Phytochemistry* **15,** 1889–1891.

Hawkes, C., and Coaker, T. H. (1979). Factors affecting the behavioral responses of the adult cabbage root fly, *Delia brassicae,* to host plant odor. *Entomol. Exp. Appl.* **25,** 45–58.

Hawkes, C., Patton, S., and Coaker, T. H. (1978). Mechanisms of host finding in adult cabbage root fly, *Delia brassicae*. *Entomol. Exp. Appl.* **24,** 419–427.

Hedin, P. A., Thompson, A. C., and Gueldner, R. C. (1976). Cotton plant and insect constituents that control boll weevil behavior and development. *Recent Adv. Phytochem.* **10,** 271–350.

Heiser, C. B., Burton, D. L., and Schilling, E. A. (1979). Biosystematic and taxonomic studies of the *Solanum nigrum* complex in eastern North America. *In* "The Biology and Taxonomy of the Solanaceae'' (J. G. Hawkes, R. N. Lester, and A. D. Skelding, eds.), pp. 513–527. Academic Press, New York.

Hsiao, T. H. (1969). Chemical basis of host selection and plant resistance in oligophagous insects. *Entomol. Exp. Appl.* **12,** 777–788.

Hsiao, T. H. (1974). Chemical influence on feeding behavior of *Leptinotarsa* beetles. *In* ''Experimental Analysis of Insect Behavior'' (L. Barton Brown, ed.), pp. 237–248. Springer–Verlag, Berlin.

Hsiao, T. H. (1978). Host plant adaptations among geographic populations of the Colorado potato beetle. *Entomol. Exp. Appl.* **24,** 437–447.

Hsiao, T. H. (1981). Ecophysiological adaptations among geographic populations of the Colorado potato beetle in North America. *In* ''Advances in Potato Pest Management'' (J. H. Lashomb and R. A. Casagrande, eds.), pp. 69–85. Hutchinson & Ross, Stroudsburg, Pennsylvania.

Hsiao, T. H., and Fraenkel, G. (1968a). Isolation of phagostimulative substances from the host plant of the Colorado potato beetle. *Ann. Entomol. Soc. Am.* **61,** 476–484.

Hsiao, T. H., and Fraenkel, G. (1968b). Selection and specificity of the Colorado potato beetle for solanaceous and non-solanaceous plants. *Ann. Entomol. Soc. Am.* **61,** 493–503.

Hsiao, T. H., and Fraenkel, G. (1968c). The influence of nutrient chemicals on the feeding behavior of the Colorado potato beetle, *Leptinotarsa decemlineata* (Coleoptera: Chrysomelidae). *Ann. Entomol. Soc. Am.* **61,** 44–54.

Hsiao, T. H., and Fraenkel, G. (1968d). The role of secondary plant substances in the food specificity of the Colorado potato beetle. *Ann. Entomol. Soc. Am.* **61,** 485–493.

Jermy, T. (1958). Untersuchungen über Aufinden und Wahl der Nahrung beim Kartoffelkäfer (*Leptinotarsa decemlineata* Say). *Entomol. Exp. Appl.* **1,** 197–208.

Jermy, T. (1961). On the nature of oligophagy in *Leptinotarsa decemlineata* Say (Coleoptera: Chrysomelidae). *Acta Zool. Acad. Sci. Hung.* **7,** 119–132.

Johnson, C. G. (1969). ''Migration and Dispersal of Insects by Flight.'' Methuen, London.

Jones, O. T., and Coaker, T. H. (1978). A basis for host plant finding in phytophagous larvae. *Entomol. Exp. Appl.* **24,** 472–484.

Kazeniac, S. J., and Hall, R. M. (1970). Flavor chemistry of volatiles. *J. Food Sci.* **35,** 519–530.

Kennedy, J. S. (1973). The plant in the life of an aphid. *Symp. R. Entomol. Soc. London* **6,** 129–140.

Kennedy, J. S. (1977a). Behaviorally discriminating assays of attractants and repellents. *In* ''Chemical Control of Insect Behavior'' (H. H. Shorey, and J. J. McKelvey, eds.), pp. 215–229. Wiley, New York.

Kennedy, J. S. (1977b). Olfactory responses to distant plants and other odor sources. *In* ''Chemical Control of Insect Behavior'' (H. H. Shorey, and J. J. McKelvey, eds.), pp. 67–91. Wiley, New York.

Kennedy, J. S., and Marsh, D. (1974). Pheromone–regulated anemotaxis in flying moths. *Science* **184,** 999–1001.

Kennedy, J. S., and Moorehouse, J. E. (1969). Laboratory observations on locust responses to wind–borne grass odour. *Entomol. Exp. Appl.* **12,** 487–503.

Kogan, M., and Goeden, R. D. (1971). Feeding and host selection behavior of *Lema trilineata daturaphila* larvae (Coleoptera: Chrysomelidae). *Ann. Entomol. Soc. Am.* **64,** 1435–1448.

Krieger, R. I., Feeny, P. P., and Wilkinson, C. F. (1971). Detoxification enzymes in the guts of caterpillars: An evolutionary answer to plant defenses? *Science* **172,** 579–581.

Ma, W.–C., and Visser, J. H. (1978). Single unit analysis of odour quality coding by the olfactory antennal receptor system of the Colorado beetle. *Entomol. Exp. Appl.* **24,** 520–533.

Matsumoto, Y. (1970). Volatile organic sulfur compounds as insect attractants with special reference to host selection. *In* "Control of Insect Behavior by Natural Products" (D. L. Wood, R. M. Silverstein, and M. Nakajima, eds), pp. 133–160. Academic Press, New York.

May, M. L. (1981). Role of body temperature and theromoregulation in the biology of the Colorado potato beetle. *In* "Advances in Potato Pest Management" (J. H. Lashomb, and R. A. Casagrande, eds.), pp. 86–104. Hutchinson & Ross, Stroudsburg, Pennsylvania.

McIndoo, N. E. (1926). An insect olfactometer. *J. Econ. Entomol.* **19,** 545–571.

Mistric, W. J., and Mitchell, E. R. (1966). Attractiveness of isolated groups of cotton plants to migrating boll weevils. *J. Econ. Entomol.* **59,** 39–41.

Mulkern, G. B. (1969). Behavioral influences on food selection in grasshoppers (Orthoptera: Acrididae). *Entomol. Exp. Appl.* **12,** 509–523.

Murray, R. D. H., Martin, A., and Stride, G. O. (1972). Identification of the volatile phagostimulants in *Solanum campylacanthum* for *Epilachna fulvosignata*. *J. Insect Physiol.* **18,** 2369–2373.

Osborne, G. O., and Hoyt, C. P. (1970). Phenolic resins as chemical attractants for males of the grass grub beetle, *Costelytra zealandica*. *Ann. Entomol. Soc. Am.* **63,** 1145–1147.

Otte, D., and Joern, A. (1977). On feeding patterns in desert grasshoppers and the evolution of specialized diets. *Proc. Acad. Nat. Sci. Philadelphia* **128,** 89–126.

Prokopy, R. J., and Haniotakis, G. E. (1975). Responses of wild and laboratory-cultured *Dacus oleae* to host plant color. *Ann. Entomol. Soc. Am.* **68,** 73–77.

Prokopy, R. J., and Haniotakis, G. E. (1976). Host detection by wild and laboratory cultured olive fruit flies. *Symp. Biol. Hung.* **16,** 209–214.

Prokopy, R. J., and Owens, E. D. (1978). Visual generalist–visual specialist phytophagous insects: Host selection behavior and application to management. *Entomol. Exp. Appl.* **24,** 609–620.

Rausher, M. D. (1978). Search image for leaf shape in a butterfly. *Science* **200,** 1071–1073.

Ricketts, H. W. (1968). "Wild Flowers of the United States," Vol. 4. McGraw–Hill, New York.

Rodriguez, J. G., Kemp, T. R., and Dabrowski, Z. T. (1976). Behavior of *Tetranychus urticae* toward essential oil mixtures from strawberry foliage. *J. Chem. Ecol.* **2,** 221–230.

Rowell, H. F. (1978). Food plant specificity in neotropical rain-forest acridids. *Entomol. Exp. Appl.* **24,** 651–662.

Saxena, K. N. (1978). Role of certain environmental factors in determining the efficiency of host plant selection by an insect. *Entomol. Exp. Appl.* **24,** 666–678.

Saxena, K. N., and Goyal, S. (1978). Host-plant relations of the citrus butterfly *Papilio demolus* L.: Orientation and ovipositional responses. *Entomol. Exp. Appl.* **24,** 1–10.

Saxena, K. N., and Saxena, R. C. (1974). Patterns of relationships between leafhoppers and plants. II. Role of sensory stimuli in orientation and feeding. *Entomol. Exp. Appl.* **17,** 493–503.

Sayo, R., and Takeo, T. (1975). Increase of *cis*-3-hexen-1-ol content in tea leaves following mechanical injury. *Phytochemistry* **14,** 181–182.

Schalk, J. M., and Stoner, A. K. (1979). Tomato production in Maryland: Effect of different densities of larvae and adults of the Colorado potato beetle. *J. Econ. Entomol.* **72,** 826–829.

Schanz, M. (1953). Der Geruchssinn des Kartoffelkäfers (*Leptinotarsa decemlineata* Say). *Z. Vgl. Physiol.* **35,** 353–379.

Schmidt–Nielsen, K. (1972). Locomotion: Energy cost of swimming, flying, and running. *Science* **177,** 222–228.

Schoonhoven, L. M. (1968). Chemosensory bases of host–selection. *Annu. Rev. Entomol.* **13,** 115–136.

Schreiber, K. (1958). Über einige Inhaltsstoffe der Solanaceen und ihre Bedeutung für die Kartoffelkäferresistenz. *Entomol. Exp. Appl.* **1,** 28–37.

Slansky, F., Jr. (1976). Phagism relationships among butterflies. *J. N.Y. Entomol. Soc.* **84,** 91–105.

Stanton, M. L. (1979). The role of chemotactile stimuli in the oviposition preference of *Colias* butterflies. *Oecologia* **39,** 79–91.

Thorsteinson, A. J. (1960). Host selection in phytophagous insects. *Annu. Rev. Entomol.* **5,** 193–218.

Tower, W. L. (1906). "An Investigation of Evolution in Chrysomelid Beetles of the Genus *Leptinotarsa*." Carnegie Inst., Washington, D.C.

van der Pers, J. C. N. (1978). Responses from olfactory receptors in three species of small ermine moths (Lepidoptera: Yponomeutidae) to plant odours. *Entomol. Exp. Appl.* **24,** 594–598.

Visser, J. H. (1976). The design of a low-speed wind tunnel as an instrument for the study of olfactory orientation in the Colorado beetle (*Leptinotarsa decemlineata*). *Entomol. Exp. Appl.* **20,** 275–288.

Visser, J. H. (1979a). Synopsis—leaf odour perception and olfactory orientation in the Colorado beetle. *In* "Olfaction in the Colorado Beetle at the Onset of Host Plant Selection" (J. H. Visser, ed.), pp. 80–85. Centre for Agricultural Publishing and Documentation, Wageningen, The Netherlands.

Visser, J. H. (1979b). Electroantennogram responses of the Colorado beetle, *Leptinotarsa decemlineata,* to plant volatiles. *Entomol. Exp. Appl.* **25,** 86–97.

Visser, J. H., and Ave, D. A. (1978). General green leaf volatiles in the orientation of the Colorado beetle, *Leptinotarsa decemlineata. Entomol. Exp. Appl.* **24,** 538–549.

Visser, J. H., and Nielsen, J. K. (1977). Specificity in the olfactory orientation of the Colorado beetle, *Leptinotarsa decemlineata. Entomol. Exp. Appl.* **21,** 14–22.

Visser, J. H., van Straten, S., and Maarse, H. (1979). Isolation and identification of volatiles in the foliage of potato, *Solanum tuberosum,* a host plant of the Colorado beetle, *Leptinotarsa decemlineata. J. Chem. Ecol.* **5,** 13–25.

Wallbank, B. E., and Wheatley, G. A. (1976). Volatile constituents from cauliflower and other crucifers. *Phytochemistry* **15,** 763–766.

Watanabe, T. (1958). Substances in mulberry leaves which attract silkworm larvae (*Bombyx mori*). *Nature (London)* **182,** 325–326.

Whalen, M. D. (1979). Speciation in *Solanum,* section *Androceras. In* "The Biology and Taxonomy of the Solanaceae" (J. G. Hawkes, R. N. Lester, and A. D. Skelding, eds.), pp. 581–596. Academic Press, New York.

Wright, R. H. (1964). After pesticides—What? *Nature (London)* **204,** 121–125.

Yamamoto, R. T., Jenkins, R. Y., and McClusky, R. K. (1969). Factors determining the selection of plants for oviposition by the tobacco hornworm *Manduca sexta. Entomol. Exp. Appl.* **12,** 504–508.

# 7

# Host-Seeking Behavior of the Gypsy Moth: The Influence of Polyphagy and Highly Apparent Host Plants*

DAVID R. LANCE



## I. INTRODUCTION

The problem of host location in generalist herbivores is somewhat different from that of more specialized feeders. Typically, specialist herbivores are adapted to utilize host plants that are relatively unpredictable in time and space. As a result, specialist feeding strategies often require host-location mechanisms that

*Preparation of this manuscript was supported, in part, by a cooperative agreement between the U.S. Department of Agriculture and the University of Massachusetts.



HERBIVOROUS INSECTS

ISBN 0-12-045580-3

are specific and that operate over relatively long distances (see Feeny *et al.*, Chapter 2; Lanier, Chapter 5; May and Ahmad, Chapter 6, this volume). Generalist herbivores utilize a larger number of plant species, and their food resources tend to be more predictable. Generalist host–location mechanisms would be expected to be less specific, bringing the insects into contact with a wide variety of plant species.

A herbivore's various host species will exhibit inter- and intraspecific differences in a number of ecologically important parameters, such as nutrient content, defensive chemistry, and phenology. As a result, an individual's choice of host plant will affect its development and reproductive success (Soo Hoo and Fraenkel, 1966a,b; Chew, 1977; Shaw and Little, 1977; Barbosa and Greenblatt, 1979). Accordingly, selection should favor herbivores that are able to locate and exploit the more suitable hosts within a given area. Taking this into consideration, host location in generalists would appear to require relatively complex mechanisms. Generalists must respond to a large number of potential hosts, but they also should be capable of selecting the superior hosts out of this battery of potentially suitable plants. Host-utilization patterns of insects in the field have been found to reflect differences in acceptability and phenology of various host species (Barbosa, 1978a; Futuyma and Wasserman, 1980). Within given locales, some generalist insects actually may function as specialists (Fox and Morrow, 1981).

The intent of this chapter is essentially twofold: (1) to examine the nature of host-location mechanisms that lead to the exploitation of what appear superficially to be diverse, predictable food resources, and (2) to look at the ability of generalist herbivores to exploit specific plant tissues that most enhance reproductive success. The gypsy moth *Lymantria dispar* L., an imported pest of forest and shade trees in the United States, will be used as a model polyphagous insect. Because of the abundance and economic importance of the species, it has been extensively studied and its biology is relatively well known.

## II. BACKGROUND BIOLOGY OF THE GYPSY MOTH

### A. Host-Plant Relationships

The host-plant range of the gypsy moth is probably in excess of 500 species. At first glance, gypsy moths may appear to be relatively nonselective feeders. In heavy infestations, they strip all of the foliage from all but a handful of tree species. They even eat plastic flagging that is tied around trees to delimit field plots. The gypsy moth therefore is an example of a highly generalist herbivore. However, foliar quality varies among the moth's various host species, and it would be advantageous for larvae to feed on the highest quality foliage within a forest stand (Hough and Pimentel, 1978; Barbosa and Greenblatt, 1979). Before we can begin to understand the degree to which gypsy moths locate the highest quality foliage within a stand, we must have a basic knowledge of the acceptability (i.e., host quality as perceived by the herbivore) and suitability (measurements of how well hervibores develop on a given plant) of some of its common host species.

In the laboratory, host acceptability can be measured by using choice tests. In choice tests, gypsy moth larvae normally prefer oaks (*Quercus* spp.) over other hosts such as red maple (*Acer rubrum*) and gray birch (*Betula populifolia*). In one series of tests, larvae preferred red oak (*Q. rubra*) over white oak (*Q. alba*) (Barbosa *et al.*, 1979), but in other tests, the larvae ate 14 times more white oak than red (Hough and Pimentel, 1978). To some degree, host preference can vary from place to place and from tree to tree.

Larval assessment of acceptability appears to involve complex mechanisms. Doskotch *et al.* (1977, 1981) examined ethanol extracts of the foliage of 190 plant species and found that the extracts can be divided into four categories: neutral extracts (82 spp.), feeding deterrents (57 spp.), feeding stimulants (30 spp.), and synergists (8 spp.). Synergists are extracts that were neutral or slightly deterrent when tested alone but, in combination with oak extract, strongly enhanced feeding. Extracts from the remaining 13 species gave variable results. A comparison of field observations with Doskotch's data indicates, as was suggested by Dethier (1973), that nonpreference can arise either from a lack of stimulants or from the presence of deterrents. Further, many different compounds can cause deterrence; 12 unique deterrents were isolated from just 3 plant species. In addition, *Catalpa speciosa* contains a number of compounds that are strongly deterrent in combination but show little or no activity when tested alone (Doskotch *et al.*, 1981). Host acceptability is apparently dependent, in part, on a wide variety of chemical cues (Dethier, 1980). Such other factors as leaf texture can also affect acceptability.

Host suitability can be measured in terms of survival, development, and fecundity of individuals that were reared on the foliage of different host-plant species. In general, the oaks are highly suitable hosts and produce large pupae and rapid development. As was the case with acceptability, the relative suitability of red versus white oak is variable. Maple and beech (*Fagus americana*) are less suitable hosts (Barbosa and Capinera, 1977; Hough and Pimentel, 1978; Barbosa and Greenblatt, 1979). Gray birch appears to be a bit of an anomoly, because it was more suitable and more readily assimilated than the oaks, yet oak was preferred over birch in choice tests (Barbosa and Greenblatt, 1979). However, herbivore-induced chemical defenses may make gray birch less suitable than oak in the field (Wallner and Walton, 1979; see also Haukioja and Niemela, 1976, 1977). Other hosts, such as pitch pine (*Pinus rigida*), are not eaten by early instars but are highly suitable for older larvae (Rossiter, 1981). Interspecific differences in host suitability appear to be significant enough to affect markedly the reproductive success of individual gypsy moths and, consequently, gypsy moth's population dynamics (Barbosa, 1978b).

## B. Life History

Adult gypsy moths are present in July and August. Female moths do not feed or fly—they merely mate, deposit an egg mass, and die. In most cases, adult females stay very close to their pupation sites. The egg masses can be found on the boles or branches of preferred hosts (e.g., red oak), on poorer hosts (e.g., red maple), or even on nonfood items (e.g., rocks, stumps, houses, and recreation vehicles). In some cases, the dispersion of egg masses is biased toward hosts that are not normally eaten by first-instar larvae (Rossiter, 1981). For example, low-level infestations have been seen in which egg masses were located primarily under bark flaps on dead trees (Lance, personal observation). The choice of pupation site (and, consequently, oviposition site) may be tied more to pressures to avoid predation than to pressures for ovipositing in the proximity of suitable foliage. Pupae that are out of the litter and in relatively sheltered sites (e.g., under bark flaps) are eaten less frequently than pupae that are in less protected areas (Bess, 1961; Campbell *et al.*, 1975a,b; Campbell and Sloan, 1976).

The eggs hatch in the spring, usually at or soon after the time of budbreak on most of the gypsy moth's host trees. Soon after hatch, the larvae enter a dispersive phase (McManus, 1973). Their dispersal occurs through ballooning; first instars drop on silk threads and are carried on the wind (Collins, 1915; Leonard, 1971). Most larvae will attempt dispersal at least once (Capinera and Barbosa, 1976). First instars have a positive phototaxis and a negative geotaxis (Wallis, 1959). Larvae climb to the tips of branches, which provide takeoff points for dispersal and where food is likely to be found. Ballooning is the moth's primary means of dispersal. Wind-borne larvae may be carried for several kilometers (Collins, 1915). However, trapping and modeling studies indicate that the vast majority of dispersal episodes result in only short-range displacements of larvae, that is, several meters (Minott, 1922; Mason and McManus, 1981). Thus a dispersal episode normally will carry a larva to a nearby tree rather than transporting it to a distant stand of trees. As females are relatively nonselective in their oviposition sites and many larvae disperse from their original host, first instars must be capable of finding suitable food.

# III. HOST LOCATION IN THE GYPSY MOTH

## A. First Instars

Host selection in insects has been considered to be the end result of two distinct processes: (1) the frequency of visitation to the different plants, which in large part is a function of the ability of the insects to locate one or more plants; and (2) the length of time that the insects spend on a given plant (Thorsteinson, 1960; Kennedy, 1965). All other things being equal, the combination of these two factors will determine host-utilization patterns in the field.

With regard to host species, ballooning is essentially a random displacement of larvae. To select a host plant, first instars would have to adjust their length of visitation—that is, a relatively strong tendency to attempt dispersal would have to be elicited when larvae perceive their host foliage to be of poor quality.

Working with caged trees, van der Linde (1971) found that first-instar gypsy moths disperse away from unacceptable foliage or unopened buds more frequently than they disperse away from preferred foliage. Since that time, larval dispersal behavior has been examined in a series of laboratory tests. Unfed first instars were placed in dispersal chambers, where they were provided with a test foliage of one of several species.

In initial tests, young apple and mature hemlock leaves were used as experimental foliage. Young apple leaves are readily accepted by first instars; on the other hand, first instars will not or cannot feed on mature hemlock foliage (Lance, personal observation). Accordingly, when hemlock was the test foliage, larvae attempted to disperse significantly more frequently than when apple was tested (Capinera and Barbosa, 1976). In a series of tests by Lance and Barbosa (1981), young leaves of red oak, white oak, gray birch, and red maple, were used. Larvae dispersed least frequently from the oaks, whereas dispersal from birch and maple was intermediate between oak and hemlock (Table I).

In outdoor tests, newly eclosed first instars were marked lightly with fluorescent dust and then released onto the boles of small potted trees (1.5 to 2.5-m

**TABLE I.** Mean Percentage of First-Instar Gypsy Moths That Attempted Dispersal in the Laboratory, in the Presence of Foliage of Several Host Species[a]

| Host | Egg Size from Which Larvae Hatched[b] | Mean Dispersal (% ± SD)[c] |
|---|---|---|
| Apple | S | 19.10 ± 11.4 |
| | M | 26.18 ± 10.1 |
| | L | 30.32 ± 13.1 |
| Red oak | S | 26.53 ± 11.8 |
| | L | 35.90 ± 13.6 |
| Red maple | S | 38.28 ± 10.7 |
| | L | 52.65 ± 12.7 |
| White oak | L | 35.88 ± 11.0 |
| Gray birch | L | 47.43 ± 6.6 |
| Eastern hemlock | S | 58.31 ± 11.4 |
| | M | 55.65 ± 11.9 |
| | L | 60.75 ± 6.3 |

[a]Data from Lance and Barbosa (1981).
[b]S, M, and L represent eggs that are $< 1.17$, 1.17–1.20 and $>1.20$ mm in diameter, respectively.
[c]Mean of four to six groups of 10 larvae each, eight observations per group over a period of 3 days.

high). The trees, three red maples and three red oaks, were held outside and their boles were banded with sticky tape to prevent larvae from crawling on and off of them. Larvae were placed on the trees in the afternoon of the day of hatch. The following evening, the trees were examined with an ultraviolet lamp, and the numbers of marked larvae on the trees were counted. More larvae remained on the oaks than on the maples (Table II). Further, there were significant differences within host species in the proportions of larvae that remained on the various trees; larvae appear to respond to intraspecific variation as well as to interspecific variation in host quality (D. R. Lance, unpublished data, 1981).

The frequency of first-instar dispersal is inversely related to the acceptability of host foliage. Consequently, preferred hosts typically carry higher larval populations than are carried by less preferred host species (Barbosa, 1978a; Lance and Barbosa, 1981). However, not all larvae will encounter highly preferred species, and, in mixed stands, many larvae feed on less preferred host plants. Host-species composition varies from stand to stand. With it, the odds of randomly encountering a preferred host plant also change. For example, if a larva finds itself on an oak, and that oak is surrounded almost exclusively by other oaks, then a dispersal episode likely would bring the larva back into contact with preferred foliage. However, if that oak is surrounded by a stand of red maples, then the larva would not be likely to reencounter preferred foliage; dispersal in this situation probably would be disadvantageous.

Parental food can influence the dispersal tendencies of first-instar gypsy moths. Larvae from different populations exhibit qualitative differences in behavior and physiology (Leonard, 1969, 1970). These differences appear to be environmentally induced rather than genetically fixed (Barbosa and Capinera, 1978). Mean egg size has been suggested as one indicator of qualitative differences among populations (Leonard, 1970; Capinera and Barbosa, 1976, 1977). The mean size of a female's eggs is influenced by the quality of the

**TABLE II.** Percentages of Marked First-Instar Gypsy Moths That Remained on Potted Trees for >30 Hr[a]

| Tree Species | Tree Number | Larvae Released | Larvae Remaining | Percentage Remaining on Tree[b] | Percentage by Host Species |
|---|---|---|---|---|---|
| Red oak | 1 | 121 | 44 | 36.4a | 31.1 |
| | 2 | 111 | 39 | 35.1a | |
| | 3 | 109 | 23 | 21.1b | |
| Red maple | 1 | 114 | 23 | 20.2b | 16.5 |
| | 2 | 94 | 6 | 6.4c | |
| | 3 | 77 | 18 | 23.4ab | |

[a]Combined results from four trials; data from D. R. Lance (unpublished data, 1981).
[b]Proportions followed by the same letter are not significantly different at $p < 0.05$.

foliage that she fed on as a larva. Feeding on highly suitable species results in a high proportion of relatively large eggs, whereas feeding on less preferred species results in higher proportions of small and medium-sized eggs. In actuality, the absolute number of small eggs in an egg mass is more or less independent of foliar quality; rather, the number of large eggs per egg mass (as well as the total number of eggs) increases with the quality of parental food (Capinera and Barbosa, 1977).

In laboratory tests, first instars from different-sized eggs varied in their tendencies to disperse. When greenhouse-grown apple leaves were used as test foliage, larvae from large eggs attempted to disperse significantly more frequently than did larvae from small eggs. Larvae from medium-sized eggs showed an intermediate tendency to disperse. When 1-year-old hemlock foliage was placed in the dispersal chambers, the frequency of dispersal was uniformly high and did not appear to vary with egg size (Table I) (Capinera and Barbosa, 1976; Lance and Barbosa, 1981). In the presence of red oak and red maple, larvae from small eggs again dispersed significantly less frequently than did larvae from large eggs (Lance and Barbosa, 1981).

In forest stands that are dominated by less preferred hosts, many individuals will feed on marginally suitable foliage; those individuals would be expected to produce a high proportion of small eggs. In the following generation, first instars would have a low tendency to disperse away from preferred food. In poor stands, preferred food is unpredictable in space. If larvae disperse away from preferred food, they would be likely to encounter only equally unsuitable or perhaps less suitable foliage. Conversely, when highly preferred hosts dominate, most larvae should have come from large eggs. These larvae would disperse readily in the presence of marginally acceptable host plants. In such stands, dispersing larvae will be likely to encounter preferred hosts. For both types of stands, intermediate dispersal should occur when host-related risks and benefits are more or less balanced. Also, unacceptable foliage always should result in dispersal, despite the composition of the surrounding stand. Laboratory tests, then, indicate that first-instar dispersal should remain appropriate in different types of forest stands. The tendency to disperse should be low in stands that provide few preferred host plants. This appears to contradict dispersal theory, which suggests that dispersal should be highest when habitats are least suitable (Southwood, 1962). However, these theories refer to interhabitat dispersal; most ballooning results in intrahabitat dispersal, and individual survival within a stand would be decreased by excessive dispersal. Also, in poorer stands, the number of dispersal attempts will be high automatically, because larvae frequently will encounter (and disperse away from) marginal and unacceptable foliage. Further, producing relative "disperser" and "nondisperser" type offspring may have important implications for dispersal and colonization by the gypsy moth (Capinera and Barbosa, 1976; Barbosa and Capinera, 1978).

The dispersal of first-instar gypsy moths has been the subject of several studies, but it remains poorly understood. For example, we do not know the actual affect of egg size on dispersal in the field. Other factors, such as weather, larval density, and larval age, also influence dispersal (Semevsky, 1971; McManus, 1973; Capinera and Barbosa, 1976). Defining the relationships among dispersal behavior, physical factors, and larval distributions may require the difficult task of following large numbers of individual larvae as they move, more or less randomly, through the forest canopy. Regardless, the gypsy moth's host-utilization patterns are explained most feasibly by the inverse relationship between foliage quality and dispersal. It is quite possible that egg size has a similar impact on larval dispersal in the field.

### B. Older Larvae

Older larvae also show qualitative differences that are dependent on the characteristics of local populations. In low-density (endemic) populations, late-instar gypsy moths feed nocturnally. They spend the day in sheltered resting sites on the branches or boles of trees or in the forest litter (Bess, 1961). In high-density populations, larvae still crawl up and down trees, but their rhythms of feeding, crawling, and resting are variable and do not appear to be strongly tied to a day–night regime.

Mark–release–recapture studies have been carried out to look at the interplant dispersal of late-instar gypsy moths (Rossiter, 1981; Lance and Barbosa, 1982; W. E. Wallner, personal communication 1982). In these studies, the boles of trees were banded with burlap to provide resting sites for the larvae (Sessions, 1895). Resting larvae were marked with fluorescent water-based paint and were placed back under the bands.

In a high-density population, 1 in every 6 recaptures was made on a different tree from the tree on which the marked larva had been released. In an endemic site, only 1 in 10 of the recaptured larvae had switched trees (Lance and Barbosa, 1982). These estimates of tree switching probably are somewhat conservative, because burlap bands on trees tend to retain some larvae that normally switch hosts (Rossiter, 1981). In some populations, late instars switch hosts frequently; older larvae also must be capable of locating host plants. In a second study in three endemic sites, 0.6–2.5% of larvae on the average switched host plants on each day (Rossiter, 1981). Host switching appears to increase with increases in either population density or tree density (Rossiter, 1981; Lance and Barbosa, 1982).

Doane and Leonard (1975) found that larvae navigate by polarized light and orient to vertical objects such as tree trunks. Thus there is a visual component to host location in gypsy moth larvae. In laboratory tests, larvae oriented more frequently to larger cardboard tree trunk models than to smaller models. Several sizes of models were used, and in general, the ratio of the numbers of orientations to one model as opposed to another was similar to the ratio of the diameters of the two models (see, e.g., Lance and Barbosa, 1982). From work on the nun moth *Lymantria monacha* L. and on *Papilio demoleus* L., it appears that this size-related effect probably was a function of the angle of the visual field that was subtended by the models (Hundertmark, 1938; Saxena and Khattar, 1977). Larvae should orient preferentially to closer trees as well as to larger ones. In the field, larvae that switch hosts normally choose a tree within several meters of their previous host (Rossiter, 1981).

Orientation to trees, and larval populations under bands on the boles of trees, appear to be related primarily to tree size (Rossiter, 1981; Lance and Barbosa, 1982). Once larvae have located a host plant, they may use silk trails to aid them in continuously relocating the same foliage and the same resting niche (McManus and Smith, 1972; Gallagher and Lanier, 1977). We do not know if chemicals play any role in the choice of tree trunks by older gypsy moth larvae. For the most part, orientation of lepidopterous larvae to olfactory stimuli has been demonstrated over only relatively short distances (Watanabe, 1958; Sutherland, 1972, 1975; Saxena and Prabha, 1975). In laboratory tests, Wallis (1959) found that oak leaves did not appear to attract gypsy moth larvae, but did act as an arrestant to larvae that contacted them. Similarly, Rafes and Ginenko (1973) suggest that host odor may play an insignificant role in the choice of tree trunks by the larvae of lepidopterous defoliators. If this is true, then host-seeking late-instar larvae also encounter different quality hosts at random, once stand composition and the physical characteristics of the trees is taken into consideration. Again, host selection primarily would be a function of the length of time that larvae spend on the various trees.

Defoliation patterns suggest that larvae feed preferentially on the more preferred host species (Baker, 1941; Campbell and Sloan, 1977). Accordingly, in Rossiter's study (1981), larvae remained on oaks significantly more faithfully than they remained on pine, and larval populations were higher on oak than on pine. In laboratory tests, larvae that had been reared on red oak were significantly more active when they were exposed to red maple foliage or to brown paper leaf models than they were when they were exposed to red oak foliage (Lance and Barbosa, 1982). When preferred foliage is not available, the larvae tend to be active and often leave their host trees. The physical structure of trees also is important; larvae more readily leave trees that do not provide preferred daytime resting sites (Rossiter, 1981; Lance and Barbosa, 1982). Another factor may contribute to observed defoliation patterns. When exposed only to less acceptable foliage, larvae consume less than they do when they are exposed to preferred foliage (Barbosa and Greenblatt, 1979; Lance and Barbosa, 1982).

### C. Influence of Host Apparency and Diet Breadth

The primary host-location mechanism of the gypsy moth is somewhat unusual among the Lepidoptera. In most Lepidoptera, ovipositing females choose the larval host plant. In the gypsy moth, host location is accomplished through larval behavior. First instars encounter potential host foliage at random, at least with respect to the distribution and physical characteristics of the plants. Depending, in part, on foliage quality, the larvae on a given plant choose either to remain or to disperse. For many Lepidoptera, interplant dispersal occurs only when food resources have been overexploited (Dethier, 1959; Myers, 1976). A high percentage of dispersing larvae likely die due to an inability to find alternate hosts (Dethier, 1959; Myers and Campbell, 1976; Chew, 1977). We might ask, why then has the gypsy moth been able to adopt interplant dispersal of larvae as a normal part of its life history? The answer apparently lies in the distribution of the gypsy moth's host plants: they are very predictable in space and, on a year-to-year basis, in time. Gypsy moths are highly polyphagous, and they also feed on highly apparent species. Apparent plants are those that are ''bound to be found'' by herbivores (Feeny, 1976). In general, apparency increases with plant size, longevity, and abundance. *Quercus,* the genus of oaks, probably is the most commonly encountered genus of trees in the hardwood forests of the northeastern United States (Harlow and Harrar, 1937).

The degree of polyphagy that a herbivore species exhibits, when combined with the spatial and temporal predictability of its host plants, will determine the overall ''encounterability'' of the food resource; that is, how frequently random encounters with plants will result in contact with suitable foliage. When encounterability of the food resource is low, host-seeking herbivores must be able to assess the quality of a very large number of potential host plants. This requires visiting a large number of plants and/or orienting only to specific stimuli, so that potential stimuli from unacceptable plants either are totally ignored or are somehow screened out. Visiting many plants or visiting widely dispersed plants requires a highly motile host-seeking stage, such as a flying adult insect.

In contrast, when suitable food is encountered readily, mechanisms that screen out unacceptable hosts may not be necessary. Further, host-seeking herbivores will not have to be capable of visiting large numbers of plants in order to locate food. When this occurs, feedback mechanisms may come into play that further increase the polyphagy of the species. First, if the food resource is readily encountered, host plants likely will be highly apparent, as is the case with forest trees and some grasses. According to current theory, apparent plants will tend to produce quantitative antiherbivore chemical defenses. These defenses act as digestibility reducers and tend to be relatively convergent (i.e., similar from species to species) when compared to the diverse toxins that are found in less apparent ephemeral plants (Feeny, 1976; Rhodes and Cates, 1976; Gilbert,

1979). In general, host-plant range may be broader for apparent plant feeders than for unapparent plant feeders (see also Futuyma, 1976).

Second, when polyphagy and host apparency are sufficiently high, an insect species may retain (or secondarily adopt) some degree of host location by the immature stages. In species in which the adult female oviposits on the host plant, immature stages often are capable of feeding and developing normally on a number of plant species that are ignored during oviposition (Wicklund, 1975; Chew, 1977). Thus host location by immatures may further promote polyphagy.

In some forest-feeding Lepidoptera, host location by immatures may occur because certain important aspects of host quality cannot be readily detected by an adult. Feeny (1970, 1976) found that foliar defensive compounds (tannins) in oak leaves tend to be present only in low quantities at the time of budbreak but subsequently build up over the following few weeks. Although tannin content is not always a reliable indicator of a tissue's digestibility (Martin and Martin, 1982), the leaves do become relatively protected from herbivores during this period. Water content, nutrient content, and leaf toughness are also involved in making the leaves progressively less suitable as they age (Scriber and Feeny, 1979).

This period of highly suitable foliage may explain why the hatch of many early-season defoliators is more or less in synchrony with budbreak of their host trees (Feeny, 1970). However, the timing of budbreak versus hatch is variable; asynchrony between the two does occur and can have severe consequences (Blais, 1957; Embree, 1965; Schneider, 1980). It is difficult to envision how the sensory apparatus of a female moth could predict, often months ahead of time, the phenological characteristics of a given tree. A female could choose a tree that was late in leafing out, and although the tree might be of a preferred host species, the female's offspring would have nothing to eat. Similarly, if the tree broke bud long before hatch, its foliage would be relatively unsuitable to the young larvae. Because the time of budbreak varies both within and among tree species, larvae that can disperse and locate new hosts may be capable of compensating for some degree of asynchrony between budbreak and hatch.

As an example, the winter moth disperses by ballooning and is one of the insects in which hatch is more or less synchronized with budbreak (Embree, 1965; Feeny, 1970). Working in an apple orchard, Holliday (1977) found that preventing oviposition on a tree did not substantially reduce that tree's subsequent larval population. Instead, what he found was a strong relationship between larval populations in the trees and the times of budbreak.

Other aspects of host-tree quality also may not be readily detectable by the adult sensory apparatus. Literature indicates that there can be substantial intraspecific differences and even intratree differences in foliage quality (Singer, 1972; Dixon, 1976; Shaw and Little, 1977; Edmunds and Alstad, 1978). Some larvae appear to be capable of responding to these differences. For example,

Schoonhoven (1977) found that late-instar gypsy moths prefer old leaves over new leaves, and leaves grown in full sunlight over leaves grown in the shade. For the gypsy moth we do not know the ecological significance of these behaviors.

## IV. HOST LOCATION IN OTHER GENERALIST HERBIVORES

Host-plant relationships and life histories vary greatly among the many species of generalist insect herbivores, and their host-location mechanisms vary accordingly. Because a broad-scale review of generalist host-location mechanisms is not feasible here, host location in two other generalist species has been chosen for review: the Japanese beetle *Popillia japonica* Newman, and the desert locust *Schistocerca gregaria* (Forskal). Unlike the gypsy moth, both of these species feed as adults. In the locust, adults and immatures utilize similar food resources (foliage of grasses and forbs). However, adults and immatures of the beetle utilize very different resources: the immatures are root feeders, whereas the adults feed on foliage or fruit. In life history and taxonomy, these two insects are quite different from the gypsy moth. However, they still must possess generalized host-location mechanisms, and they also should be capable of locating and utilizing the more suitable host plants within a habitat.

### A. The Japanese Beetle

The Japanese beetle is another example of an imported polyphagous pest. Field observations indicate that adult beetles feed on at least 295 plant species in 80 families. Extensive feeding occurs on 47 species: 14 in the Rosaceae, 5 in the Malavaceae, 4 in the Vitaceae, and 24 species spread across another 19 plant families (Fleming, 1972). The beetle larvae live in the soil and feed on the roots of a variety of grasses, ornamentals, and farm crops (Smith, 1922). Feeding behavior, along with other aspects of the insect's biology, is reviewed by Fleming (1972).

A large body of literature exists indicating that adult Japanese beetles are attracted by a variety of volatile plant compounds (Fleming, 1969; Schwartz and Hamilton, 1969; Ladd *et al.*, 1973; Ladd, 1980; Ladd and McGovern, 1980). Most of these studies were concerned with the identification of attractants that could be used for the detection, delimitation, and mass trapping of beetle populations.

Trapping studies indicate that adult Japanese beetles are attracted by a variety of unrelated plant compounds (Fleming, 1969; Schwartz and Hamilton, 1969). Several monoterpenes (especially geraniol) and phenolics (e.g., eugenol) are potent atractants. These compounds occur in preferred hosts such as rose and sassafras, among other plants (Langford *et al.*, 1943). Several phenethyl esters, most notably phenethyl acetate and phenethyl propionate, attract large numbers of beetles (Ladd *et al.*, 1973). A number of other plant compounds (e.g., some acids, alcohols, and essential oils) also act as attractants (Fleming, 1969).

When two or more attractant compounds are combined, they often act synergistically (Fleming, 1969; Ladd, 1980). Many mixtures of plant volatiles have been tested in an attempt to optimize the efficiency of traps. The more attractive mixtures invariably have contained eugenol and/or geraniol. In open areas, chemical baits may attract beetles from a distance of up to 400 m (Mehrhof and Van Leeuwen, 1930; Schwartz, 1968). However, studies of Japanese beetle attractants often have reported only the numbers of insects captured in traps and have not yielded much information on the nature of orientation mechanisms or on the relative attractiveness of trap baits as compared to host plants. Also, although compounds such as geraniol and eugenol are found in preferred host plants, their role in attracting beetles to plants has not been established.

Nevertheless, recent evidence indicates that olfaction plays a very important role in host location by adult Japanese beetles. In laboratory choice tests, intact beetles located highly preferred foliage much more frequently than less preferred foliage, but antennectomized beetles appeared to locate different types of foliage at random (Ahmad, 1982). Clearly Japanese beetles are attracted by a wide range of chemicals and chemical blends, many of which are found in a large number of plant species. The Colorado potato beetle *Leptinotarsa decemlineata* (Say) also responds to volatile chemicals that are common to a wide variety of plants. Unlike the Japanese beetle, the potato beetle does not appear to be attracted by specific compounds; rather, a blend of nonspecific chemicals in a definite ratio may be required for attraction (May and Ahmad, Chapter 6, this volume). Thus the potato beetle would be attracted to a relatively narrow range of plant species. For the Japanese beetle, variation in plant chemistry still would result in variation in attractiveness among plant species. Accordingly, the Japanese beetle should be attracted by a large number of plant species, but some host species (perhaps the more suitable?) will be more attractive than others. Currently, relationships among host attractiveness, host acceptability, and host suitability, are not well known. For example, we do not know if (or to what degree) feeding on favored versus less preferred hosts will affect parameters such as fecundity or longevity. Attractant compounds do affect acceptability as well as attractiveness. Major and Tietz (1962) found that beetles normally would not eat when they were offered leaves of *Gingko biloba*. However, when *Gingko* leaves were treated with cherry leaf extract, eugenol, or valeric acid, the beetles ate them readily without any "harmful effect." Also, compared to intact beetles, antennectomized beetles exhibit shorter feeding bouts and reduced consumption, indicating a possible integration of olfactory and gustatory cues (Ahmad, 1982). Relationships between host acceptability and tenure on the foliage have not been worked out. Sugar content also affects the acceptability of foliage (Metzger *et al.*, 1934).

Factors other than plant chemistry influence host location in the Japanese beetle. Foliage in the sunlight is more susceptible to attack than is foliage in shaded areas. Foliage in the interior of dense woodlots is attacked only rarely (Fleming, 1972). The beetles are somewhat gregarious, and plants upon which beetles are feeding are more attractive to host-seeking beetles than are plants that are devoid of beetles (see Fleming, 1972). This effect may result partially from the sex pheromone of the female beetles (Tumlinson *et al.,* 1977). In a field study, virgin females attracted 20 times more males than did traps baited with phenethyl propionate and eugenol (7:3) (Klein *et al.,* 1972). As mating often occurs on host plants (Fleming, 1972), the presence of a virgin female could attract many males to a plant.

The larvae of the Japanese beetle feed on roots, a food resource quite different from that of the adult. Larvae in the soil are quite restricted in their movements. In fallow soil, third instars traveled an average of 5 m in 29 days. When grass was present, larvae traveled an average of less than 1 m (Fleming, 1972). Larvae travel in a zigzag pattern, and the actual horizontal displacement of larvae is much less than the total distance traveled (Hawley, 1935). In areas where clumps of plants are separated by fallow soil, larvae become concentrated near the roots of the plants (Smith, 1922). Recently, Briggs and Allen (1981) demonstrated that larvae can be baited with ungerminated wheat and corn. Food plants appear to have an arrestant (and possibly attractant?) effect on the larvae. The comparative arrestant–attractant qualities of various host-plant species have not been worked out.

Because the larvae travel relatively short distances, a female's choice of oviposition site will determine the host plants available to her offspring. Female beetles oviposit in the soil among grasses or vegetable crops. Oviposition occurs most frequently in the vicinity of plants that are favored as adult food, and larval populations tend to be high in fields containing the adults' preferred host species. The phenology of adult hosts also may affect oviposition patterns (Fleming, 1972). Still, the specific host-related factors that determine oviposition site choice are not well known. Adult females may locate only host habitat for the larvae, in which case the larvae would find their own host plants. Even in areas where preferred hosts are present, many larvae travel far enough to encounter a number of different plants during the course of development. Thus the larvae apparently are capable of locating and identifying alternate host plants.

### B. The Desert Locust

The desert locust *Schistocerca gregaria* is famous in folklore for eating everything green within its path. Uvarov (1977) estimates that the host list for the locust includes over 400 species, but he warns that many of these species are eaten (or perhaps merely nibbled) only by partially starved individuals. Nevertheless, *S. gregaria*. feed readily on a large number of forbs and grasses (Williams, 1954). As in other generalist acridids, those host species that are locally most abundant tend to make up the largest proportion of an individual's diet (Paskin cited in Uvarov, 1977; for other acridid species see Bernays and Chapman, 1970a,b; Mitchell, 1975). However, the proportions of various plants in acridid diets do differ with proportions that are present in the field (Otte and Joern, 1977), and some plant species will be eaten infrequently, regardless of their abundance (Bernays and Chapman, 1970a).

To some degree, *S. gregaria* uses olfaction to locate host plants. Using a wind tunnel, Haskell *et al.* (1962) found that the locusts walked with the wind when no odor was present. The locusts turned upwind in response to the odors of grass, privet, cabbage, and some chemicals. These results were confirmed by Kennedy and Moorhouse (1970), who also noted that unfed locusts responded more readily than recently fed locusts, and that stimuli other than food odors (e.g., agitation) could elicit a switch to a positive anemotaxis. Still, using a Y-tube olfactometer, Williams (1954) was unable to demonstrate that *S. gregaria* could orient to plant odors. Dadd (1963) found that adults could discriminate among odor sources, but that newly hatched nymphs did not orient preferentially to bundles of grass over damp cotton wool (both enclosed in muslin). In contrast, Moorhouse (1971) found that locusts of all ages could respond to odors of bruised grass. Clearly odor plays some role in *S. gregaria* host location, but it is doubtful that locusts use odor to orient to individual plants within their habitat (Uvarov, 1977). Plant odors activate locusts (Kennedy and Moorhouse, 1970; Mitchell, 1981) and may function in habitat location or in the conditioning of the central nervous system (Haskell and Schoonhoven, 1969).

Orientation to visual stimuli is important in acridid host location. *S. gregaria* nymphs orient preferentially to long, straight vertical edges, to figures with complex contours, and to closer patterns over more distant ones (Wallace, 1958, 1959). Distance is not judged by binocular mechanisms or by the visual angle subtended. The nymphs sway laterally (a behavior termed *peering*) and appear to judge distance by the movement of the image across the retina. If a pattern is moved laterally while a nymph is peering, the nymph misjudges the distance to the pattern (Wallace, 1959).

Once on a host plant, *S. gregaria* assess host acceptability and leave or remain accordingly. Initially, locusts touch potential hosts with antennae and palps; normally, the plant is then bitten (Uvarov, 1977). Goodhue (1963) tested the response of *S. gregaria* to plant extracts on filter paper. Ether extracts of preferred host plants elicited a biting response, but the addition of acetone extracts was necessary before the filter paper was eaten. $R_f$ Values of the stimulants varied from plant to plant, and it was assumed that a number of different compounds could stimulate either biting or feeding. Still, in the field, acridids typically bite unacceptable plants before rejecting them (Uvarov, 1977). Starved

*Chorthippus* even attempt to bite vertical patterns on the walls of their cages (Williams, 1954), though a chemical stimulus normally may be required to elicit biting (Bernays and Chapman, 1970b). Activity levels increase when the host plant is relatively unacceptable (Bernays *et al.*, 1974). In *S. gregaria,* the final selection of food appears to depend on gustatory stimuli (Thomas, 1966). For the Acrididae, in general, host acceptability is a function of nutrient content, water content, secondary chemistry, and/or leaf texture (see Mulkern, 1979; Uvarov, 1977).

Relationships between host suitability and acceptability have not been worked out for *S. gregaria.* In *S. americana,* suitability and acceptability are correlated (Otte, 1975). Acceptability and suitability also are correlated for the acridid *Euthystira brachyptera.* However, for both *E. brachyptera* (Kaufmann, 1965) and *Melanoplus sanguinipes* (see Uvarov, 1977), a mixture of several host species provides a much more suitable diet than any single species. Further, preference for a host species can increase as the abundance of that plant species is decreased (Mitchell, 1975; preference was defined as [percentage of dry weight in diet]/[percentage of dry weight in standing crop]). Thus some acridids are adapted to utilizing a mixed diet.

Unfortunately, information on acridid–host plant interactions is notably sparse (Uvarov, 1977). An expanded knowledge of these interactions possibly could reveal trends that are contradictory to current ecological theory. For example, Joern (1979) found that the more monophagous acridid species tend to feed on more apparent plant species. This is in contrast to the predictions that have been made for the Lepidoptera and other herbivores (Feeny, 1976; Futuyma, 1976; Rhodes and Cates, 1976). In acridids, oviposition normally occurs in habitats where preferred host plants are abundant, but within those areas, females tend to choose bare patches of ground for oviposition sites (Norris, 1968). Also, most individuals feed on a number of plants during development. Immature acridids typically locate their own host plants (Mitchell, 1981), and adult acridids (ignoring gregarious phases) generally are not the strong fliers that are common among other groups such as the Lepidoptera. Typically, host-seeking acridids are not highly motile insects, and they may not be able to locate widely dispersed, unapparent plants. Thus the food resource of acridids may have to be relatively easily encountered. A high level of encounterability can be achieved either by feeding on a few highly apparent species, or by feeding on a large number of unapparent species. This is in agreement with Otte and Joern (1977), who introduce the concept of "polyphagy threshold," "an imaginary level of effort and/or time which is required to find a single suitable host species beyond which it becomes unprofitable to specialize [p. 118]." Otte and Joern contend that, as host apparency is decreased, this "time and/or effort" increases most rapidly for those insects that have limited capabilities of movement and host-plant perception. Further, differences in specialization among major insect groups (e.g.,

Lepidoptera versus Orthoptera) were attributed to differences in the efficiencies of inherent host-location mechanisms.

## V. CONCLUDING REMARKS

As a general rule, we might predict that nonspecific mechanisms would be used to bring host-seeking generalists into contact with plants. This appears to be the case with the gypsy moth (random dispersal and orientation to vertical objects), the desert locust (orientation to vertical patterns and stimulation by plant scents), and, perhaps, the Japanese beetle (which may respond to a wide range of plant volatiles). Generalized host-location mechanisms will lead to contact with foliage of variable quality. Insects that use generalized host-location mechanisms must assess foliar quality and stay on a plant or leave, depending on the acceptability of the foliage. The tenure of visits to various plants will become an important component of host selection. This strategy of host selection should be more common among generalists than it is among specialists, because specialists seldom would encounter suitable food if they contacted foliage more or less randomly. However, some generalists (e.g., the fall webworm *Hyphantria cunea*) normally complete larval development on the host plant that was chosen by the ovipositing female.

The abilities of insect herbivores to perceive host quality are limited by the insects' sensory capabilities. Although insects may perceive a large number of plant compounds and react to a *gestalt,* or overall pattern, of the compounds they perceive, their judgement of host quality cannot consider all aspects of plant chemistry (Dethier, 1973). Accordingly, some insects will refuse to feed on otherwise suitable foliage because the foliage either contains specific feeding deterrents or does not contain sufficient levels of feeding stimulants (Volkonsky, 1937; Major and Tietz, 1962; Waldbauer, 1974). This chapter has alluded to "a correlation between acceptability and suitability" as indicating that insect folivores prefer to feed on foliage that enhances reproductive fitness. Acceptability, in itself, is an important component of suitability. Thus, when examining the adaptiveness of insect–plant interactions, it is preferable to use quantitative techniques and/or techniques that partially or completely eliminate acceptability as a component of suitability (e.g., maxillectomy) (see Waldbauer, 1974; Scriber and Slansky, 1981).

The ability of insects to perceive host-plant quality also is limited by the information in the chemical message that passes from the plant to the insect. Although a number of researchers have looked at the types of information that host-seeking insects receive and the abilities of the insects to perceive that information, few if any workers have examined the information's content (i.e., how much that information can actually tell the insect about the plant) or the possible effects of limited information content on the biology of the organisms involved. For example, when an insect bite-tests a leaf, almost all of the foliar chemistry (and, therefore, information on the chemical makeup of the leaf) is available to the insect. Yet, plant volatiles make up only a fraction of the array of chemicals in a plant (Finch, 1980). If an insect uses plant volatiles as an initial step in host location, the information available to that insect is very limited, compared to the information available to an insect that encounters foliage randomly and then bite-tests. The information provided by volatile compounds may not be sufficient to allow an insect to distinguish readily some potentially suitable host plants from many other plants that would not provide suitable food. However, a number of suitable plant species may emit specific volatiles or blends of volatiles that are not emitted, for the most part, by unacceptable hosts. Thus when plant volatiles are used for oviposition site location, we might expect (as is the case) that oviposition would occur upon only a subset of potentially suitable plants, and that oviposition occasionally would occur upon plants that do not provide suitable larval food (Straatman, 1962; Wicklund, 1975; Chew, 1977; Rausher, 1979). Contact stimuli from the plant will provide additional information, but should serve primarily to restrict host-plant range further (see Feeny *et al.*, Chapter 2, this volume). Nevertheless, when the encounterability of the food resource is low, a highly motile host-seeking adult, using plant volatiles, should be able to locate host plants relatively efficiently, when compared to a much less motile immature stage. Many ecological factors, aside from host suitability, influence the host-plant relationships of insect herbivores (Otte and Joern, 1977; Gilbert, 1979; Rossiter, 1981; Futuyma, Chapter 8, this volume). In some cases, there are obvious advantages to feeding on only one or a few of the perhaps many potentially suitable host species. Certainly it is not implied here that specialists are specialists because they do not receive enough information to be generalists. However, the role of the information content of chemical messages has been largely ignored as a component in the evolution of insect–plant interactions, and it does merit some investigation.

When feasible, host location by immatures may be a more sensitive mechanism than host location by adults. Compared to insects that locate hosts via plant volatiles, insects that taste-test plants have more available information with which to judge interspecific, intraspecific, and intraplant differences in host quality (though we do not know the degree to which they are able to utilize the extra information). Further, host location by immatures can provide other options that are not available to insects in which the ovipositing female chooses the immatures' food. First instars of some forest-feeding Lepidoptera can respond to otherwise unpredictable phenological characteristics of their host plants (Holliday, 1977), and some acridids can improve the quality of their diet by feeding on several different host species (Uvarov, 1977).

A number of questions must be answered before we can begin to compare generalist immatures and generalist adults in terms of their potential for locating the most suitable foliage within a habitat. Currently, we have a restricted knowledge about the effects of intraspecific and intraplant variation on herbivores and herbivore populations. We know even less about the ability of generalist herbivores, especially immatures, to locate and exploit specific plant tissues that will provide the most suitable food. Certainly, studies on host-utilization patterns indicate that many gypsy moth larvae feed on what appears to be suboptimal foliage (Barbosa, 1978a; Lance and Barbosa, 1981).

## VI. SUMMARY

For generalist herbivores, especially those that feed on highly apparent plants, the problem of host location is different from that of specialized herbivores. In generalists such as the gypsy moth and the desert locust, random contact with foliage often will result in contact with acceptable food. Thus host location in many generalists does not require accurate, specific, long-distance orientation mechanisms. However, the host plants of generalists do vary in their suitability as food. Generalists can and do respond to this variation by preferentially feeding on the more suitable plants within a locale. It is hoped that future research will help to clarify the relationships between generalist host-location mechanisms and the nutritional and ecological requirements of the insects.

### Acknowledgments

I wish to thank S. Ahmad, T. ODell, and W. Wallner for providing unpublished manuscripts and/or unpublished data. I also wish to thank J. Elkinton for his helpful critique, and P. Barbosa for his valuable comments and his continued support.

## REFERENCES

Ahmad, S. (1982). Host location by the Japanese beetle: Evidence for a key role for olfaction in a highly polyphagous insect. *J. Exp. Zool.*, **220,** 117–120.

Baker, W. L. (1941). Effect of gypsy moth defoliation on certain forest trees. *J. For.* **39,** 1017–1022.

Barbosa, P. (1978a). Distribution of an endemic larval gypsy moth population among various tree species. *Environ. Entomol.* **7,** 526–527.

Barbosa, P. (1978b). Host plant exploitation by the gypsy moth, *Lymantria dispar* (L.). *Entomol. Exp. Appl.* **24,** 28–37.

Barbosa, P., and Capinera, J. L. (1977). The influence of food on the developmental characteristics of the gypsy moth, *Lymantria dispar* (L.). *Can. J. Zool.* **5,** 1424–1429.

Barbosa, P., and Capinera, J. L. (1978). Population quality, dispersal and numerical change in the gypsy moth, *Lymantria dispar* (L.). *Oecologia* **36,** 203–209.

Barbosa, P., and Greenblatt, J. A. (1979). Suitability, digestibility and assimilation of various host plants of the gypsy moth, *Lymantria dispar* (L.). *Oecologia* **43,** 111–119.

Barbosa, P., Greenblatt, J. A., Withers, W. A., Cranshaw, W., and Harrington, E. A. (1979). Host plant preferences and the induction of preferences in larval gypsy moths, *Lymantria dispar* (L.). *Entomol. Exp. Appl.* **26,** 180–188.

Bernays, E. A., and Chapman, R. F. (1970a). Food selection by *Chorthippus parallelus* (Zetterstedt) (Orthoptera: Acrididae) in the field. *J. Anim. Ecol.* **39,** 383–394.

Bernays, E. A., and Chapman, R. F. (1970b). Experiments to determine the basis of food selection by *Chorthippus parallelus* (Zetterstedt) (Orthoptera: Acrididae) in the field. *J. Anim. Ecol.* **39,** 761–775.

Bernays, E. A., Chapman, R. F., Horsey, J., and Leather, E. M. (1974). The inhibitory effect of seedling grasses on feeding and survival of acridids (Orthoptera). *Bull. Entomol. Res.* **64,** 413–420.

Bess, H. A. (1961). Population ecology of the gypsy moth *Porthetria dispar* L. (Lepidoptera: Lymantriidae). *Bull. Conn. Agric. Exp. Stn. New Haven,* No. 646.

Blais, J. R. (1957). Some relationships of the spruce budworm, *Choristoneura fumiferana* (Clem.) to black spruce, *Picea mariana* (Moench) Voss. *For. Chron.* **33,** 364–372.

Briggs, S. P., and Allen, W. A. (1981). Preference of *Popillia japonica* larvae to 5 baits in the laboratory. *Environ. Entomol.* **10,** 386–387.

Campbell, R. W., and Sloan, R. J. (1976). Influence of behavioral evolution on gypsy moth pupal survival in sparse populations. *Environ. Entomol.* **5,** 1211–1217.

Campbell, R. W., and Sloan, R. J. (1977). Forest stand responses to defoliation by the gypsy moth. *For. Sci. Monogr.* **19,** 1–34.

Campbell, R. W., Hubbard, D. L., and Sloan, R. J. (1975a). Patterns of gypsy moth occurrence within a sparse and numerically stable population. *Environ. Entomol.* **4,** 535–542.

Campbell, R. W., Hubbard, D. L., and Sloan, R. J. (1975b). Location of gypsy moth pupae and subsequent pupal survival in sparse, stable populations. *Environ. Entomol.* **4,** 597–600.

Capinera, J. L., and Barbosa, P. (1976). Dispersal of first instar gypsy moth larvae in relation to population quality. *Oecologia* **26,** 53–64.

Capinera, J. L., and Barbosa, P. (1977). Influence of natural diets and larval density on gypsy moth, *Lymantria dispar* (Lepidoptera: Orgyiidae), egg mass characteristics. *Can. Entomol.* **109,** 1313–1318.

Chew, F. S. (1977). Coevolution of pierid butterflies and their cruciferous food plants. II. The distribution of potential food plants. *Evolution* **31,** 568–579.

Collins, C. W. (1915). Dispersion of gypsy-moth larvae by the wind. *U.S. Dep. Agric. Bur. Entomol. Bull.,* No. 273.

Dadd, R. H. (1963). Feeding behavior and nutrition in grasshoppers and locusts. *Adv. Insect Physiol.* **1,** 47–109.

Dethier, V. G. (1959). Food-plant distribution and density and larval dispersal as factors affecting insect populations. *Can. Entomol.* **91,** 581–596.

Dethier, V. G. (1973). Electrophysiological studies of gustation in lepidopterous larvae. II. Taste spectra in relation to food-plant discrimination. *J. Comp. Physiol.* **82,** 103–134.

Dethier, V. G. (1980). Evolution of receptor sensitivity to secondary plant substances with special reference to deterrents. *Am. Nat.* **115,** 45–66.

Dixon, A. F. G. (1976). Timing of egg hatch and viability of the sycamore aphid, *Drepanosiphum platanoidis* (Schr.), at budburst of sycamore, *Acer pseudoplatanus* L. *J. Anim. Ecol.* **45,** 593–603.

Doane, C. C., and Leonard, D. E. (1975). Orientation and dispersal of late stage larvae of *Porthetria dispar* (Lepidoptera: Lymantriidae). *Can. Entomol.* **107,** 1333–1338.

Doskotch, R. W., ODell, T. M., and Girard, L. (1981). Phytochemicals and feeding behavior of gypsy moth larvae. *U.S. Dep. Agric. Tech., Bull.* No. 1584.

Doskotch, R. W., ODell, T. M., and Godwin, P. A. (1977). Feeding responses of gypsy moth larvae, *Lymantria dispar,* to extracts of plant leaves. *Environ. Entomol.* **6,** 563–566.

Edmunds, G. F., and Alstad, D. N. (1978). Coevolution in insect herbivores and conifers. *Science (Washington, D.C.)* **199,** 941–945.

Embree, D. G. (1965). The population dynamics of the winter moth in Nova Scotia, 1945–1962. *Mem. Entomol. Soc. Can.,* No. 46. 57.

Feeny, P. (1970). Seasonal changes in oak leaf tannins and nutrients as a cause of spring feeding by winter moth caterpillars. *Ecology* **51,** 565–581.

Feeny, P. (1976). Plant apparency and chemical defense. *Recent Adv. Phytochem.* **10,** 1–40.

Finch, S. (1980). Chemical attraction to plant-feeding insects to plants. *Appl. Biol.* **5,** 67–143.

Fleming, W. E. (1969). Attractants for the Japanese beetle. *U.S. Dep. Agric. Tech. Bull.,* No. 1399.

Fleming, W. E. (1972). Biology of the Japanese beetle. *U.S. Dep. Agric. Tech. Bull.,* No. 1449.

Fox, L. R., and Morrow, D. A. (1981). Specialization: Species property or local phenomenon. *Science* **211,** 887–893.

Futuyma, D. J. (1976). Food plant specialization and environmental predictability in Lepidoptera. *Am. Nat.* **110,** 285–292.

Futuyma, D. J., and Wasserman, S. S. (1980). Resource concentration and herbivory in oak forests. *Science* **210,** 920–921.

Gallagher, E. M., and Lanier, G. N. (1977). Trail following behavior in the gypsy moth caterpillar *Porthetria dispar* (L.), (Lepidoptera, Lymantriidae). *J. N.Y. Entomol. Soc.* **85,** 174–175.

Gilbert, L. E. (1979). Development of theory in the analysis of insect–plant interactions. *In* "Analysis of Ecological Systems (D. J. Horn, G. R. Stairs, and R. D. Mitchell, eds.), pp. 117–154. Ohio State Univ. Press, Columbus.

Goodhue, D. (1963). Feeding stimulants required by a polyphagous insect, *Schistocerca gregaria. Nature* (*London*) **197,** 405–406.

Harlow, W. M., and Harrar, E. S. (1937). "Textbook of Dendrology." McGraw–Hill, New York.

Haskell, P. T., and Schoonhoven, L. M. (1969). The function of certain mouth part receptors in relation to feeding in *Schistocera gregaria* and *Locusta migratoria migratorioides. Entomol. Exp. Appl.* **12,** 423–440.

Haskell, P. T., Paskin, M. W., and Moorhouse, J. E. (1962). Laboratory observations on factors affecting the movements of hoppers of the desert locust. *J. Insect Physiol.* **8,** 53–78.

Haukioja, E., and Niemela, P. (1976). Does birch defend itself actively against herbivores? *Rep. Kevo Subarctic Res. Stn.* **13,** 44–47.

Haukioja, E., and Niemela, P. (1977). Retarded growth of a geometrid larva after mechanical damage to leaves of its host tree. *Ann. Zool. Fenn.* **14,** 48–52.

Hawley, I. M. (1935). Horizontal movement of larvae of the Japanese beetle in field plots. *J. Econ. Entomol.* **28,** 656.

Holliday, N. J. (1977). Population ecology of the winter moth (*Operophtera brumata*) on apple in relation to larval dispersal and time of budburst. *J. Appl. Ecol.* **14,** 803–813.

Hough, J. H., and Pimentel, D. (1978). Influence of host foliage on development, survival, and fecundity of the gypsy moth. *Environ. Entomol.* **7,** 97–101.

Hundertmark, A. (1938). Das formenuntescheidungs-vermogen Eiraupen der nonne (*Lymantria monacha L.*). *Z. Vgl. Physiol.* **24,** 563–582.

Joern, A. (1979). Feeding patterns in grasshoppers (Orthoptera: Acrididae): Factors influencing diet specialization. *Oecologia* **38,** 325–347.

Kaufmann, T. (1965). Biological studies on some Bavarian Acridoidea (Orthoptera), with special reference to their feeding habits. *Ann. Entomol. Soc. Am.* **58,** 791–801.

Kennedy, J. S. (1965). Mechanisms of host plant selection. *Ann. Appl. Biol.* **56,** 317–320.

Kennedy, J. S., and Moorhouse, J. E. (1970). Laboratory observations on locust responses to windborne grass odour. *Entomol. Exp. Appl.* **12,** 487–503.

Klein, M. G., Ladd, T. L., Jr., and Lawrence, K. O. (1972). A field comparison of lures for Japanese beetles: Unmated females vs. phenethyl propionate + eugenol (7:3). *Environ. Entomol.* **1,** 397–399.

Ladd, T. L., Jr. (1980). Japanese beetle: Enhancement of lures by eugenol and caproic acid. *J. Econ. Entomol.* **73,** 718–720.

Ladd, T. L., and McGovern, T. P. (1980). Japanese beetle: A superior attractant,phenethyl propionate + eugenol + geraniol 3:7:3. *J. Econ. Entomol.* **73,** 689–691.

Ladd, T. L., Jr., McGovern, T. P., Beroza, M., Townshend, B. G., Klein, M. G., and Lawrence, K. O. (1973). Japanese beetles: Phenethyl esters as attractants. *J. Econ. Entomol.* **66,** 369–370.

Lance, D. R., and Barbosa, P. (1981). Host tree influences on the dispersal of first instar gypsy moths, *Lymantria dispar* (L.). *Ecol. Entomol.* **6,** 411–416.

Lance, D. R., and Barbosa, P. (1982). Host tree influences on the dispersal of late instar gypsy moths, *Lymantria dispar* (L.). *Oikos* **38,** 1–7.

Langford, G. S., Muma, M. H., and Cory, E. N. (1943). The attractiveness of certain plant constituents to the Japanese beetle. *J. Econ. Entomol.* **36,** 248–252.

Leonard, D. E. (1969). Intrinsic factors causing qualitative changes in populations of the gypsy moth. *Proc. Entomol. Soc. Ont.* **100,** 195–199.

Leonard, D. E. (1970). Intrinsic factors causing qualitative changes in populations of *Porthetria dispar* (Lepidoptera: Lymantriidae). *Can. Entomol.* **102,** 239–249.

Leonard, D. E. (1971). Air-borne dispersal of the gypsy moth and its influence on concepts of control. *J. Econ. Entomol.* **64,** 638–641.

Major, R. T., and Tietz, H. J. (1962). Modification of the resistance of *Gingko biloba* leaves to attack by Japanese beetles. *J. Econ. Entomol.* **55,** 272.

Martin, J. S. and Martin, M. M. (1982). Tannin assays in ecological studies: Lack of correlation between phenolics, proanthocyanidins, and protein-precipitating constituents in mature foliage of six oak species. *Oecologia* **54,** 205–211.

Mason, C. J., and McManus, M. L. (1981). Larval dispersal of the gypsy moth. *USDA Tech. Bull.,* No. 1584.

McManus, M. L. (1973). The role of behavior in the dispersal of newly hatched gypsy moth larvae. *U.S.D.A. For. Serv. Res. Pap. NE,* No. 267.

McManus, M. L., and Smith, H. R. (1972). Importance of silk trails in the diel behavior of late instars of the gypsy moth. *Environ. Entomol.* **1,** 793–795.

Mehrhof, F. E., and Van Leeuwen, E. R. (1930). An electric trap for killing Japanese beetles. *J. Econ. Entomol.* **23,** 275–278.

Metzger, F. W., Vander Meullen, P. A., and Mell, C. W. (1934). The relation of sugar content and odor of clarified extracts of plants to their susceptibility to attack by Japanese beetle. *J. Agric. Res.* **49,** 1001–1008.

Minott, C. W. (1922). The Gypsy moth on cranberry bogs. *U.S. Dep. Agric. Bur. Entomol. Bull.,* No. 1093.

Mitchell, J. E. (1975). Variation in food preferences of three grasshopper species (Acrididae: Orthoptera) as a function of food availability. *Am. Midl. Nat.* **94,** 267–283.

Mitchell, R. (1981). Insect behavior, resource exploitation, and fitness. *Annu. Rev. Entomol.* **26,** 183–211.

Moorhouse, J. E. (1971). Experimental analysis of the locomotor behavior of *Schistocerca gregaria* induced by odour. *J. Insect Physiol.* **17,** 913–920.

Mulkern, G. B. (1967). Food selection by grasshoppers. *Annu. Rev. Entomol.* **12,** 59–78.

Myers, J. H. (1976). Distribution and dispersal in populations capable of resource depletion: A simulation model. *Oecologia* **23,** 255–269.

Myers, J. H., and Campbell, B. J. (1976). Distribution and dispersal in populations capable of resource depletion: A field study on cinnabar moth. *Oecologia* **24,** 7–20.

Norris, M. T. (1968). Laboratory experiments on oviposition responses of the desert locust, *Schistocerca gregaria* (Forskal). *Anti-Locust Bull.* **43,** 1–47.

Otte, D. (1975). Plant preference and plant succession: A consideration of evolution of plant preference in *Schistocerca*. *Oecologia* **18,** 129–141.

Otte, D., and Joern, A. (1977). On feeding patterns in desert grasshoppers and the evolution of specialized diets. *Proc. Acad. Nat. Sci. Philadelphia* **128,** 89–126.

Rafes, P. M., and Ginenko, Yu. I. (1973). The survival of leaf-eating caterpillars (Lepidoptera) as related to their behavior. *Entomol. Rev.* (*Engl. Transl.*) **52,** 204–211. [Engl. Transl. of Entomologicheskoe Obozrenie].

Rausher, M. D. (1979). Larval habitat suitability and oviposition preference in three related butterflies. *Ecology* **60,** 503–511.

Rhodes, D. F., and Cates, R. G. (1976). Toward a general theory of plant antiherbivore chemistry. *Recent Adv. Phytochem.* **10,** 168–213.

Rossiter, M. C. (1981). ''Factors Contributing to Host Range Extension in the Gypsy Moth, *Lymantria dispar.*'' Ph.D. Dissertation, State Univ. of New York, Stony Brook.

Saxena, K. N., and Khattar, P. (1977). Orientation of *Papilio demoleus* larvae in relation to size, distance, and combination pattern of visual stimuli. *J. Insect Physiol.* **23,** 1421–1428.

Saxena, K. N., and Prabha, P. (1975). Relationship between olfactory sensilla of *Papilio demoleus* L. larvae and their orientation responses to different odours. *J. Entomol. Ser. A. Physiol. Behav.* **50,** 119–126.

Schneider, J. C. (1980). The role of parthenogenesis and female aptery in microgeographic, ecological adaptation in the fall cankerworm, *Alsophila pometaria* Harris (Lepidoptera: Geometridae). *Ecology* **61,** 1082–1090.

Schoonhoven, L. M. (1977). Feeding behaviour in phytophagous insects: On the complexity of the stimulus situation. *Colloq. Int. C. N. R. S.,* No. 265, 391–398.

Schwartz, P. H. (1968). Distribution of released Japanese beetles in a grid of traps. *J. Econ. Entomol.* **61,** 423–426.

Schwartz, P. H., and Hamilton, D. W. (1969). Attractants for the Japanese beetle. *J. Econ. Entomol.* **62,** 516–517.

Scriber, J. M., and Feeny, P. (1979). Growth of herbivorous caterpillars in relation to feeding specialization and to the growth form of their plants. *Ecology* **60,** 829–850.

Scriber, J. M., and Slansky, F., Jr. (1981). The nutritional ecology of immature insects. *Annu. Rev. Entomol.* **26,** 183–211.

Semevsky, F. N. (1971). Optimization of larval behavior in *Porthetria dispar* under their distribution in the grown [*sic*]. *Zh. Obshch. Biol.* **3,** 312–316.

Sessions, W. R. (1895). Extermination of the gypsy moth. *Rep. Mass. State Board Agric.,* No. 4.

Shaw, C. G., and Little, C. H. A. (1977). Natural variation in balsam fir foliar components of dietary importance to the spruce budworm. *Can. J. For. Res.* **7,** 47–53.

Singer, M. C. (1972). Complex components of habitat suitability within a butterfly colony. *Science* **176,** 75–77.

Smith, L. B. (1922). Larval food habits of the Japanese beetle (*Popillia japonica* Newm.). *J. Econ. Entomol.* **15,** 305–310.

Soo Hoo, C. F., and Fraenkel, G. (1966a). The selection of food plants in a polyphagous insect, *Prodenia eridania* (Cramer). *J. Insect Physiol.* **12,** 693–709.

Soo Hoo, C. F., and Fraenkel, G. (1966b). The consumption, digestion, and utilization of food plants by a polyphagous insect, *Prodenia eridania* (Cramer). *J. Insect Physiol.* **12,** 711–730.

Southwood, T. R. E. (1962). Migration of terrestrial arthropods in relation to habitat. *Biol. Rev. Cambridge Philos. Soc.* **37,** 171–214.

Straatman, R. (1962). Notes on certain Lepidoptera ovipositing on plants which are toxic to their larvae. *J. Lepid. Soc.* **16,** 99–103.

Sutherland, O. W. (1972). The attraction of newly hatched codling moth (*Laspeyresia pomonella*) larvae to apple. *Entomol. Exp. Appl.* **15,** 481–487.

Sutherland, O. W. (1975). Response of newly hatched codling moth larvae (*Laspeyresia pomonella*) to water vapour. *Entomol. Exp. Appl.* **18,** 389–390.

Thomas, J. G. (1966). The sense organs on the mouthparts of the desert locust (*Schistocera gregaria*). *J. Zool.* **148,** 420–448.

Thorsteinson, A. J. (1960). Host selection in phytophagous insects. *Annu. Rev. Entomol.* **5,** 193–218.

Tumlinson, J. H., Klein, M. G., Doolittle, R. E., Ladd, T. L., and Proveaux, A. T. (1977). Identification of the female Japanese beetle sex pheromone: Inhibition of male response by an enantiomer. *Science* **197,** 789–792.

Uvarov, B. (1977). "Grasshoppers and Locusts: A Handbook of General Acridology," Vol. 2. Centre Overseas Pest Research, London.

van der Linde, R. J. (1971). The sailing flight of the gypsy moth (*Lymantria dispar* L.) and the effect of the food plant on this phenomenon. *Z. Angew. Entomol.* **67,** 316–324.

Volkonsky, M. A. (1937). Sur l'action acridifuge des extraits de feuilles de *Melia azedarach. Arch. Inst. Pasteur Alger.* **15,** 427–432.

Waldbauer, G. P. (1974). Consumption, digestion, and utilization of solanaceous and non-solanaceous plants by larvae of the tobacco hornworm, *Protoparce sexta* (Johan.) (Lepidoptera: Sphingidae). *Entomol. Exp. Appl.* **17,** 253–269.

Wallace, G. K. (1958). Some experiments on form perception in nymphs of the desert locust, *Schistocerca gregaria* Forskal. *J. Exp. Biol.* **35,** 765–775.

Wallace, G. K. (1959). Visual scanning in the desert locust *Schistocerca gregaria* Forskal. *J. Exp. Biol.* **36,** 512–525.

Wallis, R. C. (1959). Factors affecting larval migration of the gypsy moth. *Entomol. News* **70,** 235–240.

Wallner, W. E., and Walton, G. S. (1979). Host defoliation: A possible determinant of gypsy moth population quality. *Ann. Entomol. Soc. Am.* **72,** 62–67.

Watanabe, T. (1958). Substances in mulberry leaves which attract silkworm larvae (*Bombyx mori*). *Nature* (*London*) **182,** 325–326.

Wicklund, C. (1975). The evolutionary relationship between adult oviposition preferences and larval host plant range in *Papilio machaon* L. *Oecologia* **18,** 185–197.

Williams, L. H. (1954). The feeding habits and food preferences of the Acrididae and the factors which determine them. *Trans. R. Entomol. Soc. London* **105,** 423–454.

# PART IV

# Evolutionary Aspects of Host Selection

# 8

# Selective Factors in the Evolution of Host Choice by Phytophagous Insects*

DOUGLAS J. FUTUYMA



## I. INTRODUCTION

This chapter addresses the question that the Editor posed in convening the contributors to this volume: Is there a unifying theme amid the diversity of responses of phytophagous insects to plants? The behavioral and neurophysiological studies of phytophagous insects reported in these chapters and elsewhere (Schoonhover and Dethier, 1966; Schoonhover, 1973; Dethier, 1980) present a bewildering diversity of idiosyncratic phenomena: insects range in diet from enormously polyphagous to extraordinarily host specific; they respond to plants by using one or several of their sensory modalities, including vision, olfaction, and taste and other forms of contact chemoreception; their behavioral responses

*Support during the preparation of this chapter was provided by the National Science Foundation (DEB 76-20232). This is Contribution No. 399 in Ecology and Evolution from the State University of New York at Stony Brook.



HERBIVOROUS INSECTS

ISBN 0-12-045580-3

to plants are sometimes explicable in terms of the responses of chemosensory cells and sometimes in terms of the integration of chemosensory (and other) input by the central nervous system. A simple organizing theme that will subsume the responses of hundreds of thousands of species of insects to hundreds of thousands of species of plants thus cannot be provided. Instead, however, an attempt will be made, primarily from the viewpoint of evolutionary biology, at least to suggest the limits of generalization that may ultimately prove possible.

There are two major approaches to biology, and it is possible to seek a unifying theme for our subject in either. Functional biology asks how organisms work—What mechanisms are responsible for biological phenomena? Evolutionary biology asks how organisms came to be the way they are—What are the historical reasons for the evolution of biological mechanisms? Thus we may seek either a unifying mechanistic theme or a unifying evolutionary theme, to organize our knowledge of insect responses to plants.

## II. ALTERNATIVE PHYSIOLOGICAL MECHANISMS

It may be asked, for example, Why do insect species use one sensory modality or another to acquire food, and why do they process sensory information in one way or another to generate feeding or oviposition behavior? The answer inevitably will have an immediate evolutionary, or historical, component: in part, the sensory modalities used must be constrained by the insect's anatomy, which is in large part phylogenetically determined. Thus orders and families of insects differ taxonomically in their capacity for sustained flight, which must influence their ability to rely on olfactory or visual information obtained at a distance, and in the anatomical distribution and kinds of chemoreceptors. The presence of tarsal chemoreceptors in some groups but not others is more likely to be determined by phylogeny than by needs of a particular species.

In large part, however, whatever sensory modality is used, the behavioral response of an insect to a plant is determined by the sensitivity of sensory cells (e.g., chemoreceptors) to plant stimuli, and the processing of sensory information by the central nervous system (CNS). A given plant compound, or physical stimulus, is commonly a repellent to one species and an attractant to another. Nonspecific plant compounds, such as sugars, are generally stimulatory to feeding or oviposition, but are typically necessary rather than sufficient stimuli: they cannot account for the host specificity of specialized insects. Host-specific compounds such as the glucosinolates of crucifers may well have evolved as plant defenses, and be repellent to a wide variety of insects, but are reinterpreted as attractants by species that specialize on crucifers. The genetic changes in behavior that enable an insect to be attracted to a formerly repellent plant presumably can be of several kinds: (1) a loss of the ability of the sensory system to perceive the repellent; (2) perception of the compound, but reinterpretation of the stimulus by the CNS so that it is no longer repellent; and (3) interpretation of the compound, or of another compound in the same plant, as a necessary attractant. Changes (1) or (2) will result in an expansion of the diet to include a previously rejected plant; change (3) results in specificity for the plant as a new host.

Some of these changes may be illustrated (Table I) by the responses of the sibling species of small ermine moths in the *Yponomeuta padellus* complex (Yponomeutidae) to their hosts in Europe (Herrebout *et al.*, 1976; Gerrits-Heybroek *et al.*, 1978; van der Pers, 1978; van Drongelen, 1980; van Drongelen and van Loon, 1980). For several reasons, including the utilization of *Euonymus* (Celastraceae) by species of *Yponomeuta* outside the *padellus* complex, *Euonymus* is thought to be the ancestral host. The relevant species for our present purposes are (1) *Y. cagnagellus,* which feeds on *Euonymus* only; (2) *Y. evonymellus,* which despite its name feeds on *Prunus* (Rosaceae), although it will accept *Euonymus* when deprived of a choice; and (3) *Y. malinellus,* which feeds on *Malus* (Rosaceae) and will not accept *Euonymus*. The lateral sensillum of *Y. cagnagellus* shows an electrophysiological response to dulcitol, the distinctive compound of the species' host, and the larvae are behaviorally attracted to dulcitol. *Yponomeuta evonymellus* responds to dulcitol in the same way both behaviorally and electrophysiologically, but prefers *Prunus;* thus dulcitol is not sufficient to elicit feeding. *Yponomeuta malinellus* shows neither an electrophysiological nor a behavioral response to dulcitol. All the species respond electrophysiologically to prunasin, but whereas *Y. cagnagellus* is repelled by it, *Y. evonymellus* shows no behavioral response. Thus *Y. evonymellus* could have

**TABLE I.** Responses to Plants of Sibling Species of *Yponomeuta*[a,b]

| Response | Natural Host/Host Compound | | |
|---|---|---|---|
| | *Euonymus* dulcitol *Y. cagnagellus* | *Prunus* prunasin *Y. evonymellus* | *Malus* phloridzin *Y. malinellus* |
| Dulcitol | | | |
| Behavioral | + | + | 0 |
| Sensory | + | + | − |
| Prunasin | | | |
| Behavioral | − | 0 | ? |
| Sensory | + | + | + |
| Phloridzin | | | |
| Behavioral | − | − | 0 |
| Sensory | + | + | − |

[a]Information compiled from references cited in text.
[b]+, positive response; −, negative (sensory) or avoidance (behavioral) response; 0, no response.

adopted *Prunus* as a host by losing its CNS interpretation of prunasin as a deterrent, and acquiring a positive response to unidentified properties of *Prunus*. Both *Y. cagnagellus* and *Y. evonymellus* respond electrophysiologically to phloridzin, the distinctive compound in *Malus,* and are repelled by it. *Yponomeuta malinellus,* which feeds on *Malus,* does not respond electrophysiologically to phloridzin and is not repelled by it. Thus *Y. malinellus* could have evolved onto *Malus* by losing its sensory sensitivity to both the attractant dulcitol in *Euonymus* and to the repellent phloridzin in *Malus* and by acquiring a positive response to other features of *Malus*. Crosses between *Y. cagnagellus* and *Y. malinellus* give results compatible with the hypothesis that the differences in electrophysiological responses to the compounds could be due to very simple genetic changes.

The difference in behavioral response of *Y. cagnagellus* and *Y. evonymellus* to prunasin appears to be a difference in CNS interpretation of sensory information; the difference between *Y. malinellus* on the one hand and *Y. cagnagellus* and *Y. evonymellus* on the other, in their behavioral response to phloridzin, appears to be mediated by a change in the peripheral sensory system. This example suggests that evolutionary host shifts may come about by genetically minor alterations either of primary sensory input or of CNS processing of information: even within the confines of this small species group, both kinds of changes seem to have occurred. Thus it seems unlikely that any broad mechanistic generalizations will emerge about the sensory and central nervous responses of diverse insects to their hosts.

## III. MODES OF EVOLUTIONARY EXPLANATION

Evolution, however, is the chief unifying principle of biology, and there is every reason to hope that it can provide unifying themes for insect–plant interactions as it has for other biological problems. Evolutionary biology as a source of explanation has, however, two persistent themes that are to some degree antithetical as sources of explanation: *evolutionary history* (including phylogeny) and *adaptation*. They generate quite different kinds of explanations.

Historical evolutionary studies interpret the present properties of organisms as the product of their phylogenetic past. These properties may or may not have emerged as adaptations to past conditions, but they are not necessarily explicable by reference to present environmental conditions. The possible adaptive responses that a species may mount to environmental challenges are severely constrained by the properties bequeathed by its ancestors. Thus scale insects may be forever condemned to an essentially sessile existence, cimicids to flightlessness, and heliconiine butterflies to ovipositing on Passifloraceae. Each of the functional systems of an organism has, in this view, severe developmental constraints on its evolutionary malleability (Gould and Lewontin, 1979). Organisms are never as optimally adapted as an engineer might design them to be. Moreover, the differences among species are not necessarily special, optimal adaptations to the peculiar environment that each species occupies; they may be alternative adaptive solutions to the same environmental problem, and the solution adopted by a species is a consequence of the historical accidents of environment and genetic composition suffered by its remote ancestors. Whether *Yponomeuta evonymellus* adapted to prunasin by turning off its sensory response to this repellent or by altering its central nervous interpretation of this response is most likely a matter of historical accident.

In contrast, adaptation as an evolutionary explanation seeks answers in the current ecology of a species. In recent years, especially in evolutionary ecology and behavior, workers have turned to *optimization theory* to predict or explain characteristics as more or less ideal responses to current or recent environmental conditions. The "adaptationist programme," as Gould and Lewontin (1979) term it in their critical essay, assumes adaptation and assumes that developmental pathways and the availability of genetic variation do not pose serious constraints on the direction or rate of evolution. It is at a loss to explain nonadaptive features, such as the possession of rudimentary secondary sexual characteristics by the "wrong" sex (e.g., mammae in men and wattles in hens).

Both traditions of evolutionary explanation are likely necessary to account for the diversity of responses of insects to plants, and this chapter explores them both by asking why it is that every insect species feeds on some plants but not others. This is not quite the same as asking why a plant is not attacked by every species of insect that encounters it. Plants unquestionably have features that exclude some insects. The question here, however, is whether the exclusion of plants from an insect's diet is purely because of an inability to adapt to plant defenses, or whether specialization may stem from other evolutionary reasons as well.

## IV. PHYLOGENETIC CONSIDERATIONS

That phylogenetic considerations can have paramount importance is obvious from the taxonomic distribution of feeding habits in insects, which has long been known to entomologists and was brought to general attention by Ehrlich and Raven (1964). There is a strong phylogenetic component to host specificity in many groups such as the butterflies (Ehrlich and Raven, 1964) and aphids (Eastop, 1973). It is clear that most Coliadinae feed on legumes, Pierini on crucifers, and Troidini on Aristolochiaceae because their ancestors became specialized on these plant groups. There is little point in asking why the monarch butterfly does not feed on crucifers even when milkweeds are rare—it presumably does not recognize them as possible food, and relegates them to the category "nonmilkweed" with most of the rest of the green world.

The phylogenetic effect emphasized by Ehrlich and Raven (1964) should not be overstated, however, for in almost every insect group, at least some related species feed on phylogenetically and chemically very dissimilar hosts. Within the butterflies, for example, this is especially true in the Lycaenidae (Gilbert, 1979). In the beetle genus *Epilachna,* related species feed on legumes and cucurbits; in *Rhagoletis* (Diptera: Tephritidae), sibling species specialize on rosaceous, cornaceous, or ericaceous fruits (Bush, 1966). These and many other examples suggest, together with other lines of evidence, that many plant barriers are easily overcome in insect evolution. In some cases, the switch to an unrelated plant is actually a switch to a plant that shares similar chemical stimuli with the ancestral host, and so is readily explicable (Ehrlich and Raven, 1964). Often, however, there is no evident chemical commonality between the hosts of related insects, and we must ask why a species should switch to a particular novel host rather than another. In what way, if any, is the host onto which the species evolves better than others not chosen?

## V. COST–BENEFIT CONSIDERATIONS

The common answer is cast in terms of plant defenses: that adaptation to the specific toxins is required for utilization of a particular plant, and that the insect achieves this adaptation at the cost of effectively dealing with the chemistry of other plants. Thus specialization is the product of the "cost" of polyphagy; as MacArthur (1972) put it, "a jack of all trades is master of none." To be sure, ecological factors other than plant chemistry are sometimes invoked to explain the advantage of host specificity, but interspecific competition and escape from host-associated enemies (Gilbert and Singer, 1975; Price *et al.*, 1980) play a lesser role in the plant–insect literature than toxic or otherwise deleterious plant compounds (see e.g., most of the chapters in Rosenthal and Janzen, 1979).

Unquestionably, many plant compounds are toxic to unadapted insects, and the species that use these plants have adaptations for dealing with their toxic qualities (e.g., Self *et al.*, 1964; Feeny, 1976; Bernays *et al.*, 1980; Berenbaum and Feeny, 1981). It seems likely, however, that many, perhaps most, insects can quickly evolve the ability to deal with many, perhaps most, defensive chemicals *if* they are not deterred at the behavioral level, and so expose themselves to selection for resistance to the toxin. The first step in a change of host utilization must be a behavioral one; following this, resistance to the physiological barriers to growth and survival can probably quickly evolve. Thus the toxins peculiar to a group of plants are probably less of an evolutionary barrier to their utilization than is generally supposed. Several lines of evidence can lead to this conclusion.

First, the major detoxification mechanisms of insects are ubiquitous among insect species, and most of them, such as the mixed-function oxidase (MFO) system, can effectively detoxify an extraordinary range of chemical compounds (Agosin and Perry, 1974; Dauterman and Hodgson, 1978; Brattsten, 1979). Few species-specific detoxification systems (i.e., adapted to particular plant toxins) seem to have been reported (a possible example of such, however, is given by Rosenthal *et al.*, 1978). Moreover, insects have abundantly demonstrated their ability to adapt to an enormous range of foreign compounds, namely, insecticides. Adaptation to insecticides often occurs within a few generations, and is commonly achieved not by the evolution of new detoxifying enzymes, but by increases in the level of activity of enzymes such as the MFOs (Brattsten, 1979). High MFO activity may be the primary mechanism by which polyphagous insects adapt to the broad range of phytochemicals they encounter (Krieger *et al.*, 1971). The induction of high MFO activity and other detoxifying enzymes by ingestion of foreign compounds (Brattsten *et al.*, 1977) shows that insects have the machinery for quantitative regulation of these enzymes; genetic fixation of higher activity would be easily achieved. This may entail some energetic cost (Schoonhoven and Meerman, 1978), but this cost is probably minor.

Second, there are many reports (Waldbauer and Fraenkel, 1961; Smiley, 1978) that specialized insects can grow and survive quite normally when reared on plants very different from their natural hosts. In some cases this is revealed by depriving them of their ability to differentiate among plants, and so inducing them to feed on plants they would otherwise reject.

Third, there is some evidence that specialized insects assimilate their food more efficiently than nonspecialists, but the generality of this effect is equivocal. That the cost of detoxification may not be very high is illustrated by Scriber's finding (1978) of equal efficiency of assimilation in southern armyworms fed on cyanogenic and acyanogenic morphs of *Lotus corniculatus*. Scriber (1979) reported that the specialist *Papilio troilus* (spicebush swallowtail) grew more efficiently than the polyphagous *Papilio glaucus* (tiger swallowtail) on spicebush, the host of *P. troilus;* but Futuyma and Wasserman (1981) found no difference in efficiency of the generalist *Malacosoma disstria* (forest tent caterpillar) and the oligophagous *Malacosoma americanum* when both were reared on the major host of *M. americanum*, black cherry (which is also one of the many hosts used by *M. disstria*). Our laboratory has also been comparing parthenogenetic genotypes of the fall cankerworm *Alsophila pometaria* (Geometridae), some of which are most prevalent in stands of oak (*Quercus* spp.) and others in stands of red maple (*Acer rubrum*). The phenological correspondence between the time of hatching of these genotypes and the time of budbreak of their associated hosts may be the major reason for the genetic differentiation (Mitter *et al.*, 1979). Oak-associated genotypes tend not to accept maple foliage, whereas a maple-associated genotype does; but in preliminary tests, we have found no evidence of differences in assimilation efficiency that indicate adaptation of these genotypes to their respective hosts (D. J. Futuyma, I. van Noordwijk, and R. P. Cort, unpublished).

Fourth, the ability of insects to adapt readily to new hosts is attested by the rapid origin of "biotypes" adapted to previously resistant strains of crops. Adapted genotypes of many species of aphids have arisen in response to planting new species and strains of many kinds of crops, and at least in some cases the adaptation may consist largely of the willingness to feed on plants that are simply not accepted by nonadapted genotypes (Nielson and Don, 1974).

Fifth, one might expect polymorphism for the utilization of different host species to be common in insect populations if specialization were achieved at the cost of adaptation to other hosts. A stable "multiple niche" polymorphism can theoretically occur if each of several genotypes is better adapted to a particular resource than to those used by other genotypes (Levene, 1953; Maynard Smith, 1966; Bulmer, 1974). Hence an abundance of cases of host races, or polymorphism for host utilization, might imply that fitness trade-offs are common. This notion, moreover, is integral to the idea that insects may speciate sympatrically by adapting to different hosts: genotypes that are attracted to different hosts would be subject to selection for reproductive isolation if each is less fit on the other's host (Bush, 1974). Indeed Bush (1975) and others have proposed that sympatric speciation by host-race formation may account for the prevalence of host-specific insects.

## VI. GENETIC VARIATION IN PHYTOPHAGOUS INSECTS

However, there are few, if any, well-documented cases of sympatric host races (Futuyma and Mayer, 1980; Jaenike, 1981). Many putative cases of host races have proved to be complexes of host-specific sibling species that are reproductively isolated (Herrebout *et al.*, 1976, Jaenike and Selander, 1980; Guttman *et al.*, 1981). The famous case of the apple, hawthorn, and cherry host races of *Rhagoletis pomonella* (Bush, 1974) is insufficiently documented (Futuyma and Mayer, 1980; Paterson, 1981). An electrophoretic survey of several polyphagous species of geometrids provided no evidence of host-associated polymorphism (Mitter and Futuyma, 1979). The best-documented cases of sympatric host races appear to be in: (1) *Papilio demodocus*, in which differently colored larvae are found on rutaceous and umbelliferous plants, probably because of predation (Clarke *et al.*, 1963); (2) scale insects adapted to individual ponderosa pine trees (Edmunds and Alstad, 1978), a situation fostered by their low vagility and multiple successive generations on a single tree; and (3) aphids and the fall cankerworm (Mitter *et al.*, 1979), in which parthenogenesis presumably permits the retention of specific gene combinations required for adaptation to different hosts. The paucity of host-associated polymorphisms in sexual species suggests that either fitness trade-offs are uncommon or recombination prevents the persistence of host-adapted genotypes. It also argues against the prevalence of sympatric speciation.

Nevertheless, populations do harbor genetic variation that can be mobilized for adaptation to new hosts, but such adaptive divergence generally occurs, I believe, among allopatric populations, in response to the prevalence of different hosts in different places. The rapid evolution of host affiliation in crop pests, such as aphids, the brown planthopper of rice (Pathak, 1970; Panda, 1979), and the codling moth—which has shifted from apple to walnut and plum in California (Phillips and Barnes, 1975)—has occurred where the new host or resistant strain is abundantly planted, and the original host species or strain is rare. Hsiao (1978) has found that populations of Colorado potato beetle from Arizona, where *Solanum elaeagnifolium* is more abundant than the more usual host *S. rostratum,* shows a greater preference for and adaptation to *S. elaeagnifolium* than other populations of the beetle (see also May and Ahmad, Chapter 6, this volume). Local populations of the butterfly *Euphydryas editha* have also diverged in host preference (Singer, 1971) and in physiological adaptation (Rausher, 1982). Faced with only a single host, an insect population may evolve a preference for it, to the exclusion of previously utilized host species. In laboratory selection experiments on the bruchid beetle *Callosobruchus maculatus,* Wasserman and Futuyma (1981) found that populations reared for 11 generations only on pigeon peas developed almost as pronounced a preference for this host as did populations that were artificially selected for choosing pigeon peas over azuki beans.

## VII. EVOLUTION OF HOST-SEEKING BEHAVIOR

Initially, the barriers to utilization of a new host are twofold: the plant does not provide the necessary stimuli to attract the insect, and it has chemical or physical properties that reduce the fitness of those individuals that are attracted to it. The requirement that individuals arise that are genetically capable of surmounting both these obstacles is one consideration that makes it unlikely that host-specific species arise by sympatric speciation (Futuyma and Mayer, 1980). In a local population faced with an abundance of an inferior host and rarity of its normal host, however, selection favors attraction to the new host, even if its defensive chemistry reduces survival. Once the behavioral barrier is overcome, selection will favor physiological adaptation to the toxic or other inimical properties of the plant.

The idea that an insect population should evolve a preference for a locally abundant host is hardly revolutionary. It does, however, raise a serious question: Why should a population that initially prefers Host A over Host B develop an exclusive preference for B when it is abundant, and lose its response to A? Why should it not merely expand its diet, and acquire greater potential polyphagy?

No doubt this is exactly what sometimes happens. For example, the maple-associated genotype of *Alsophila* has not (yet, at least) lost its responsiveness to oak. For at least two major reasons, the development of a positive response to a locally abundant host may entail the loss of response to other host species. Both explanations involve adaptation, but both also entail the supposition of constraints and are thus not part of the idealist adaptationist program.

First, the possibility of making mistakes can select for oligophagy. In an important paper that has been much neglected, Levins and MacArthur (1969) pointed out that if the neurological machinery of an insect is capable of only a limited capacity for decisions, the animal may be unable to distinguish between palatable plants that support growth and toxic plants that do not. An ovipositing female then has reduced fitness if she lays eggs on the unpalatable plant. Eisner and Grant (1981) have suggested that the chemical sign stimuli of palatable plants might well resemble those of toxic or otherwise unsuitable species. Levins and MacArthur show that especially if a chemically identifiable palatable plant is abundant, a female insect with limited search time may accrue highest fitness by ovipositing only on this plant, to the exclusion of those palatable species that cannot be distinguished from the unpalatable one.

For example, Rodman and Chew (1980) have found that females of the butterfly *Pieris napi macdunnoughi* lay eggs both on a native crucifer (*Descurainia richardsonii*) that supports larval growth, and on two alien crucifers (*Thlaspi arvense* and *Chorispora tenella*) that are lethal to the larvae, but that share certain glucosinolates with the *Descurainia*. In such circumstances, selection would favor alleles that (1) conferred tolerance to the toxic crucifers, or (2) enabled discrimination between the *Descurainia* and the toxic aliens, or (3) led to abandonment of all three species in favor of the other crucifers that the butterfly also uses successfully. If alleles that program alternative (3) are more prevalent than those that confer the other kinds of adaptation, the species will abandon both *Descurainia* and the two aliens, and there will be no selection for alternatives (1) or (2) if the genetic variations permitting such adaptations should arise later. In this case, the species will have evolved a narrower diet. Whether or not the species retains *Descurainia* in the diet therefore depends on the historical accident of which kinds of genetic variations are present at a particular point in the species' history. It is therefore relevant to ask whether populations tend to harbor more selectable genetic variation in behavioral responses or in the physiological ability to detoxify novel phytochemicals. Wasserman and Futuyma (1981) found that host preference responded to selection in bruchid beetles reared on various mixtures of bean species, whereas adaptation of physiological tolerance to the inferior host did not. These experiments suggest that behavioral responses may be more genetically malleable. Clearly, however, both behavioral and physiological evolution has occurred in crop pests, and it would be premature to postulate whether one or the other kind of trait is usually more genetically variable.

A second possible related reason for the evolution of a specialized preference for a locally common host is suggested by the existence of "search images" in both vertebrates and at least one butterfly (Rausher, 1980). Vertebrate predators searching for food, and females of *Battus philenor* searching for oviposition sites, preferentially attack the most common of several alternative prey, and in doing so increase their rate of "harvest" (Dawkins, 1971; Murton, 1971; Rausher, 1980). At least in chickens, the ability to perceive a particular kind of item is enhanced by focusing attention on it, and ignoring alternatives (Dawkins, 1971). Thus sensory systems, or the ability of the central nervous system to process sensory information, are constrained in ways that favor specialization. In principle, therefore, a genotype with a genetically fixed search image for a particular kind of prey item or host could find prey faster than a genotype without such a search image.

The possible advantage of specialization is easily modeled. Following Holling (1959) and Charnov (1976), let $L_i$ be the number of host plants of Species i encountered per unit search time and $h_i$ be the time spent recognizing and laying an egg on a host plant of Species i, once it is encountered. A female butterfly spends a total time $T_s$ searching for hosts for oviposition. The total time spent in oviposition is $L_1h_1T_s + L_2h_2T_s$ if she accepts both Host Species 1 and 2. The total time $T$, then, is searching time plus handling time, or $T = T_s + T_s(L_1h_1 + L_2h_2) = T_s(1 + L_1h_1 + L_2h_2)$. The number of eggs laid is $L_1T_s + L_2T_s$ for a polyphagous genotype that accepts both hosts, and is $L_1T_s$ for a monophagous genotype that accepts only Host 1. Assuming that the female has an effectively unlimited store of eggs, that the risk of mortality favors maximization of oviposition rate, and that there is no intraspecific competition (i.e., that hosts are not rejected because they already bear eggs), the fitter genotype will be the one with the higher value of $G$ = eggs laid/time. For the polyphagous genotype, $G_p = (L_1 + L_2)T_s/[T_s(1 + L_1h_1 + L_2h_2)]$. For the monophagous genotype, $G_m = L_1T_s/T_s(1 + L_1h_1)$. Now assume that $h_1 = h_2 = h_p$ for the polyphagous genotype (handling time is equal for both plants), and that $h_1 = h_m$ for the monophagous genotype, where $h_m < h_p$ because the monophagous form, with a fixed search image for Host 1, recognizes its host faster. Then

$$G_p = \frac{(L_1 + L_2)}{1 + h_p(L_I + L_2)}$$

$$G_m = L_1(1 + h_mL_1)$$

If we set $h_m = ch_p (c < 1)$ and set the fitness of the polyphagous genotype $W_p = G_p = 1$, the relative fitness of the monophagous genotype is

$$W_{\mathrm{m}} = \frac{G_{\mathrm{m}}}{G_{\mathrm{p}}} = \frac{1 + h_{\mathrm{p}}(L_1 + L_2)}{1 + ch_{\mathrm{p}}(L_1 + L_2) + L_2/L_1}$$

Clearly, as $L_2$ declines relative to $L_1$ (i.e., as the abundance of Host 2 decreases), the denominator decreases faster than the numerator because of the factor $c$.

Figure 1 gives some numerical results of these equations. The smaller the value of $c$ (i.e., the more efficient the monophagous genotype is in recognizing its host, compared to the polyphagous genotype), the greater is $W_{\mathrm{m}}$, the relative fitness of the specialized genotype. As the abundance of the specialist's host increases (i.e., as $L$ decreases in Fig. 1), $W_{\mathrm{m}}$ comes to exceed 1; that is, it has higher fitness than the polyphagous genotype. If host recognition is subject to the constraints that the phenomenon of search image implies, it can be advantageous for a population faced with an abundance of one host to lose the capacity to recognize ancestral, but locally rare, hosts as oviposition sites. Again, the accidents of history—in this case the spatial distribution and abundance of host plants—can determine the course of adaptation.

The search image model presented here carries some implications that may constitute limited tests of the hypothesis. First, it supposes that oviposition rate

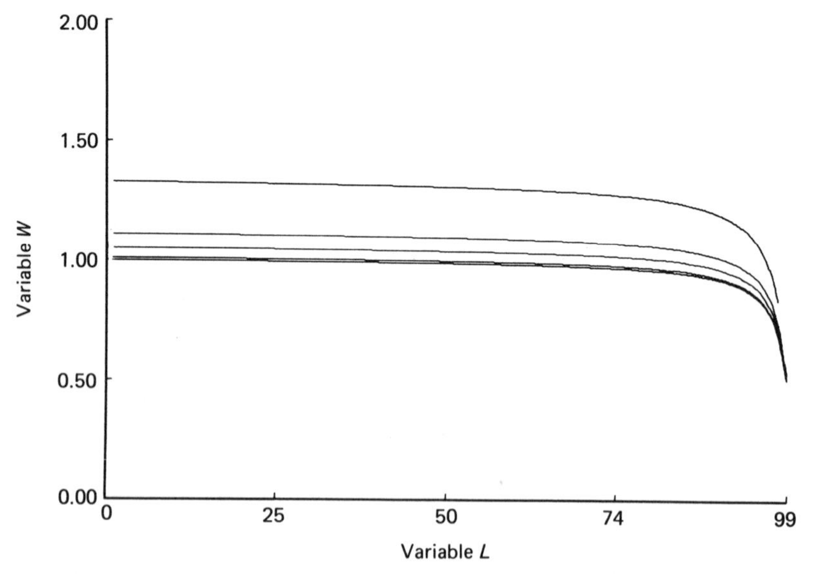


**Fig. 1.** The fitness $W$ of a monophagous genotype that searches for Host Species 1, relative to that of a polyphagous genotype that searches for Hosts 1 and 2. $L$ is the abundance of Host 2 and corresponds to $L_2$ in the equation $W_{\mathrm{m}} = [1 + h(L_1 + L_2)]/[1 + ch(L_1 + L_2) + L_2/L_1]$, where $L_1 + L_2 = 100$. The recognition time per encountered plant is $h = 1$ for the polyphagous genotype and $ch$ for the monophagous genotype. Higher curves represent lower values of $c$ ($c = 0.75, 0.90, 0.95, 0.99, 1.00$), that is, more efficient recognition.

should be increased by search imagelike behavior; Rausher (1980) has shown that it is possible to investigate this possibility and to find evidence of such facilitation. Second, it predicts, as optimal foraging theory generally does (Krebs, 1973; Pyke *et al.*, 1977), that specialization should be favored in populations that are sparse relative to the abundance of hosts; as competition increases, an advantage accrues to genotypes that will accept un-utilized and rarer hosts. (This prediction cannot be tested, however, by comparing densities of monophagous and polyphagous species: a species that has already evolved hostspecialization may become abundant enough to experience competition. We would predict, however, that more polyphagous species should show increase catholicity of oviposition preference by individual females as population density increases.) Third, host-specific species should evolve more often from other host-specific species than from polyphagous ancestors. The term $L_2$ in the equations just given is less likely to be small relative to $L_1$ (the abundance of Host 1) if $L_2$ represents a multitude of plant species accepted by a generalized insect, than if it represents a single alternative host accepted by an oligophagous insect.

It is critical, in this interpretation, that the shift to specialization on a new host plant be accompanied or followed by the acquisition of reproductive isolation. Once a population has been behaviorally entrained to a plant and ceases to exchange genes with populations occupying other plants, it is possible for selection to favor the development of specialized detoxifying systems, so that the species becomes physiologically attuned to the idiosyncrasies of its host. To the extent that a population, or a complex of populations that exchange genes at a high rate is polyphagous, there will be correspondingly little selection for adaptation to the properties of any one host species. Adaptation to one host can sometimes confer cross-adaptation to other hosts (Gould, 1979), but cross-adaptation to chemically very different host plants is probably less likely. Thus plants that are chemically highly distinctive, in either repellent or toxic properties, will tend to support only specialized insects. It is argued here, however, that the ability to use distinctive, toxic hosts is contingent on the prior evolution of behavioral specialization, which brings the population into an ecological context in which selection for physiological adaptation can operate. Thus plant toxins unquestionably pose barriers to insects, but they are barriers that probably are often easily overcome, once behavioral specialization evolves. If this is so, we would predict that the adaptation of, say, flea beetles and pierine butterflies to crucifers was not due to any special preadaptation to mustard oil glycosides; that a polyphagous insect such as the southern armyworm or a specialist such as the Mexican bean beetle could adapt to these compounds just as readily, if they were forced by circumstances to make a behavioral switch. This is perhaps overstating the case, but this viewpoint is consistent with the observation that related species of insects often specialize on chemically very different, unrelated plants.

Dethier (1947, 1970) has long argued that the key to understanding host-plant specialization lies in behavioral adaptation. Some other authors (e.g., Feeny, 1976; Rhoades, 1979) have laid greater stress on the physiological barriers that plant chemicals present to insects. The importance of behavior has been clearly stressed here, but the intention is not to denigrate the impact that plant toxins have played in insect evolution. In many instances a coevolution of behavioral and physiological adaptations must be involved; in other cases, a more purely behavioral explanation may suffice. An attempt has been made here, however, to counter the strict optimalist thinking that is prevalent in many areas of evolutionary ecology, by stressing that an insect's diet is not necessarily the best possible diet it could have chosen; that though the evolution of diet may be guided by natural selection, it is not necessarily guided along the best possible path.

## VIII. CONCLUDING REMARKS

What, then, can be said about the development of an unifying theory of host-plant affiliations in insects? If we seek a powerfully predictive theory of what the hosts for undescribed or little-known insects should be, the outlook cannot be very optimistic. As Lewontin (1967) has pointed out, even the same set of environmental variations can have entirely different genetic outcomes, depending on the temporal order of the environmental changes. As in historical processes generally, the outcome of evolution is determined by unknowable past sequences of events. Even if we hold to the belief that all genetic changes are governed by natural selection—and there is no reason for such a belief—the evolution of feeding responses can follow utterly different paths, depending on the history of local environments and the order in which favorable genetic variations arise. To be sure, phylogenetic affinity may serve as a loose predictor of host-plant affiliation, and optimization theory may often predict an insect's breadth of diet, based on information about the abundance, phenology, and temporal variation in the availability of hosts. But the diversity of the responses of the hundreds of thousands of species of phytophagous insects is not amenable to easy adaptive generalizations divorced from evolutionary history.

## IX. SUMMARY

Cost–benefit models and optimization theory are unlikely to provide a powerfully predictive theory of the evolution of host preference in phytophagous insects. This is partly because of phylogenetic constraints, and partly because of the indeterminacy of genetic change: the initial conditions of both genetic variation and local ecological circumstances can set populations off on any of several paths of adaptive evolution. Examples of alternate evolutionary paths that might not be predicted from a strict optimization approach are described, at the levels of both physiological mechanisms and host-seeking behavior.

Host-plant specialization in insects often entails specialized physiological mechanisms, but there is little evidence that the evolution of such mechanisms is responsible for specialization and precludes the maintenance of a broader diet. Rather, specialization is posited to evolve first at a behavioral level, perhaps in response to the local abundance of a particular plant species. Behavioral specialization then sets the stage for the evolution of physiological adaptations such as detoxification that are specific to the secondary compounds peculiar to the host.

## Acknowledgments

I am grateful to Robert Armstrong for guidance in the formulation of the equations, and to Scott Ferson for obtaining the computer printout. The members of the insect ecology discussion group at Stony Brook have provided invaluable criticism and suggestions.

## REFERENCES

Agosin, M., and Perry, A. S. (1974). Microsomal mixed-function oxidases. *In* "The Physiology of Insecta" (M. Rockstein, ed.), Vol. 5, pp. 538–596. Academic Press, New York.

Berenbaum, M., and Feeny, P. (1981). Toxicity of angular furanocoumarins to swallowtail butterflies: Escalation of a coevolutionary arms race? *Science* **212,** 927–929.

Bernays, E. A., Chamberlain, D., and McCarthy, P. (1980). The differential effects of ingested tannic acid on different species of Acridoidea. *Entomol. Exp. Appl.* **28,** 158–166.

Brattsten, L. B. (1979). Biochemical defense mechanisms in herbivores against plant allelochemicals. *In* "Herbivores: Their Interaction with Secondary Plant Metabolites" (G. A. Rosenthal, and D. H. Janzen, eds.), pp. 200–270. Academic Press, New York.

Brattsten, L. B., Wilkinson, C. F., and Eisner, T. (1977). Herbivore–plant interactions: Mixed-function oxidases and secondary plant substances. *Science* **196,** 1349–1352.

Bulmer, M. G. (1974). Density-dependent selection and character displacement. *Am. Nat.* **108,** 45–58.

Bush, G. L. (1966). The taxonomy, cytology, and evolution of the genus *Rhagoletis* in North America (Diptera, Tephritidae). *Bull. Mus. Comp. Zool.* **134,** 431–562.

Bush, G. L. (1974). The mechanism of sympatric host race formation in the true fruit flies (Tephritidae). *In* "Genetic Mechanisms of Speciation in Insects" (M. J. D. White, ed.), pp. 3–23. Aust. & N.Z. Book Co., Sydney.

Bush, G. L. (1975). Modes of animal speciation. *Annu. Rev. Ecol. Syst.* **6,** 339–364.

Charnov, E. L. (1976). Optimal foraging: Attack strategy of a mantid. *Am. Nat.* **110,** 141–151.

Clarke, C. A., Dickson, C. G. C., and Sheppard, P. M. (1963). Larval color pattern in *Papilio demodocus*. *Evolution* **17,** 130–137.

Dauterman, W. C., and Hodgson, E. (1978). Detoxication mechanisms in insects. *In* "Biochemistry of Insects" (M. Rockstein, ed.), pp. 541–577. Academic Press, New York.

Dawkins, M. (1971). Perceptual changes in chicks: Another look at the "search image" concept. *Anim. Behav.* **19,** 566–574.

Dethier, V. G. (1947). "Chemical Insect Attractants and Repellents." McGraw-Hill (Blakiston), New York.

Dethier, V. G. (1970). Chemical interactions between plants and insects. *In* "Chemical Ecology" (E. Sondheimer, and J. B. Simeone, eds.), pp. 83–102. Academic Press, New York.

Dethier, V. G. (1980). Evolution of receptor sensitivity to secondary plant substances with special references to deterrents. *Am. Nat.* **115,** 45–66.

Eastop, V. F. (1973). Deductions from the present day host plants of aphids and related insects. *Symp. R. Entomol. Soc. London* **6,** 157–178.

Edmunds, G. F., Jr., and Alstad, D. N. (1978). Coevolution in insect herbivores and conifers. *Science* **199,** 941–945.

Ehrlich, P. R., and Raven, P. H. (1964). Butterflies and plants: A study in coevolution. *Evolution* **18,** 586–608.

Eisner, T., and Grant, R. P. (1981). Toxicity, odor aversion, and "olfactory aposematism." *Science* **213,** 476.

Feeny, P. P. (1976). Plant apparency and chemical defense. *Recent Adv. Phytochem.* **10,** 1–40.

Futuyma, D. J., and Mayer, G. C. (1980). Non-allopatric speciation in animals. *Syst. Zool.* **29,** 254–271.

Futuyma, D. J., and Wasserman, S. S. (1981). Food plant specialization and feeding efficiency in the tent caterpillars *Malacosoma disstria* Hübner and *M. americanum* (Fabricius). *Entomol. Exp. Appl.* **30,** 106–110.

Gerrits–Heybroek, E. M., Herrebout, W. M., Ulenberg, S. A., and Wiebes, J. T. (1978). Host plant preferences of five species of small ermine moths (Lepidoptera, Yponomeutidae). *Entomol. Exp. Appl.* **24,** 360–368.

Gilbert, L. E. (1979). Development of theory in the analysis of insect–plant interactions. *In* "Analysis of Ecological Systems" (D. Horn, R. Mitchell, and G. Stairs, eds.), pp. 117–154. Ohio State Univ. Press, Columbus.

Gilbert, L. E., and Singer, M. C. (1975). Butterfly ecology. *Annu. Rev. Ecol. Syst.* **6,** 365–397.

Gould, F. (1979). Rapid host range evolution in a population of the phytophagous mite *Tetranychus urticae* Koch. *Evolution* **33,** 791–802.

Gould, S. J., and Lewontin, R. C. (1979). The spandrels of San Marco and the Panglossian paradigm: A critique of the adaptationist programme. *Proc. R. Soc. London Ser. B.* **205,** 147–164.

Guttman, S. I., Wood, T. K., and Karlin, A. A. (1981). Genetic differentiation along host plant lines in the sympatric *Enchenopa binotata* Say complex (Homoptera: Membracidae). *Evolution* **35,** 205–217.

Herrebout, W. M., Kuijten, P. J., and Wiebes, J. T. (1976). Small ermine moths of the genus *Yponomeuta* and their host relationships (Lepidoptera, Yponomeutidae). *In* "The Host-Plant in Relation to Insect Behavior and Reproduction" (T. Jermy, ed.), pp. 91–94. Plenum, New York.

Holling, C. S. (1959). Some characteristics of simple types of predation and parasitism. *Can. Entomol.* **91,** 385–398.

Hsiao, T. H.(1978). Host plant adaptations among geographic populations of the Colorado potato beetle. *Entomol. Exp. Appl.* **24,** 237–247.

Jaenike, J. (1981). Criteria for ascertaining the existence of host races. *Am. Nat.* **117,** 830–834.

Jaenike, J., and Selander, R. K. (1980). On the question of host races in the fall webworm *Hyphantria cunea. Entomol. Exp. Appl.* **27,** 31–37.

Krebs, J. R. (1973). Behavioral aspects of predation. *In* "Perspectives in Ecology" (P. P. G. Bateson, and P. H. Klopfer, eds.), pp. 73–111. Plenum, New York.

Krieger, R. I., Feeny, P. P., and Wilkinson, C. F. (1971). Detoxification enzymes in the guts of caterpillars: An evolutionary answer to plant defenses? *Science* **172,** 579–581.

Levene, H. (1953). Genetic equilibrium when more than one ecological niche is available. *Am. Nat.* **87,** 331–333.

Levins, R., and MacArthur, R. (1969). An hypothesis to explain the incidence of monophagy. *Ecology* **50,** 910–911.

Lewontin, R. C. (1967). The principle of historicity in evolution. *Wistar Inst. Symp. Monogr.* **5,** 81–94.

MacArthur, R. H. (1972). "Geographical Ecology." Harper & Row, New York.

Maynard Smith, J. (1966). Sympatric speciation. *Am. Nat.* **100,** 637–650.

Mitter, C., and Futuyma, D. J. (1979). Population genetic consequences of feeding habits in some forest Lepidoptera. *Genetics* **92,** 1005–1021.

Mitter, C., Futuyma, D. J., Schneider, J. C., and Hare, J. D. (1979). Genetic variation and host plant relations in a parthenogenetic moth. *Evolution* **33,** 777–790.

Murton, L. K. (1971). The significance of a specific search image in the feeding behaviour of the wood-pigeon. *Behaviour* **40,** 10–42.

Nielson, M. W., and Don, H. (1974). Probing behavior of biotypes of the spotted alfalfa aphid on resistance and susceptible alfalfa clones. *Entomol. Exp. Appl.* **17,** 477–486.

Panda, N. (1979). "Principles of Host-Plant Resistance to Insect Pests." Allanheld, Osmun, and Co., New York.

Paterson, H. E. H. (1981). The continuing search for the unknown and unknowable: A critique of contemporary ideas on speciation. *S. Afr. J. Sci.* **77,** 113–119.

Pathak, M. D. (1970). Genetics of plants in pest management. *In* "Concepts of Pest Management" (R. L. Rabb, and F. E. Guthrie, eds.), pp. 138–157. North Carolina State Univ. Press, Raleigh.

Phillips, P. A., and Barnes, M. M. (1975). Host race formation among sympatric apple, walnut, and plum populations of the codling moth, *Laspeyresia pomonella. Ann. Entomol. Soc. Am.* **68,** 1053–1060.

Price, P. W., Bouton, C. E., Gross, P., McPheron, B. A., Thompson, J. N., and Weis, A. E. (1980). Interactions among three trophic levels: Influence of plants on interactions between insect herbivores and natural enemies. *Annu. Rev. Ecol. Syst.* **11,** 41–65.

Pyke, G. H., Pulliam, H. R., and Charnov, E. L. (1977). Optimal foraging: A selective review of theory and tests. *Q. Rev. Biol.* **52,** 137–154.

Rausher, M. D. (1980). Host abundance, juvenile survival, and oviposition preference in *Battus philenor. Evolution* **34,** 342–355.

Rausher, M. D. (1982). Population differentiation in *Euphydryas editha* butterflies: Larval adaptations to different hosts. *Evolution* **36,** 581–590.

Rhoades, D. F. (1979). Evolution of plant chemical defense against herbivores. *In* "Herbivores: Their Interaction with Secondary Plant Metabolites" (G. A. Rosenthal, and D. H. Janzen, eds.), pp. 3–54. Academic Press, New York.

Rodman, J. E., and Chew, F. S. (1980). Phytochemical correlates of herbivory in a community of native and naturalized Cruciferae. *Biochem. Syst. Ecol.* **8,** 43–50.

Rosenthal, G. A., and D. H. Janzen (eds.) (1979). "Herbivores: Their Interaction with Secondary Plant Metabolites." Academic Press, New York.

Rosenthal, G. A., Dahlman, D. L., and Janzen, D. H. (1978). L-canavanine detoxification: A seed predator's biochemical mechanism. *Science* **202,** 528–529.

Schoonhoven, L. M. (1973). Plant recognition by lepidopterous larvae. *Symp. R. Entomol. Soc. London* **6,** 87–99.

Schoonhoven, L. M., and Dethier, V. G. (1966). Sensory aspects of host-plant discrimination by lepidopterous larvae. *Arch. Neerl. Zool.* **16,** 497–530.

Schoonhoven, L. M., and Meerman, J. (1978). Metabolic cost of changes in diet and neutralization of allelochemics. *Entomol. Exp. Appl.* **24,** 489–493.

Scriber, J. M. (1978). Cyanogenic glycosides in *Lotus corniculatus:* Their effect upon growth, energy budget, and nitrogen utilization of the southern armyworm *Spodoptera eridania*. *Oecologia* **34,** 143–155.

Scriber, J. M. (1979). The effects of sequentially switching foodplants upon biomass and nitrogen utilization by polyphagous and stenophagous *Papilio* larvae. *Entomol. Exp. Appl.* **25,** 203–215.

Self, L. S., Guthrie, E. F., and Hodgson, E. (1964). Adaptation of tobacco hornworm to the ingestion of nicotine. *J. Insect Physiol.* **10,** 907–914.

Singer, M. C. (1971). Evolution of food–plant preference in the butterfly *Euphydryas editha*. *Evolution* **25,** 383–389.

Smiley, J. (1978). Plant chemistry and the evolution of host specificity: New evidence from *Heliconius* and *Passiflora*. *Science* **201,** 745–747.

van der Pers, J. C. N. (1978). Responses from olfactory receptors in females of three species of small ermine moths (Lepidoptera: Yponomeutidae) to plant odours. *Entomol. Exp. Appl.* **24,** 394–398.

van Drongelen, W. (1980). Behavioural responses of two small ermine moth species (Lepidoptera: Yponomeutidae) to plant constituents. *Entomol. Exp. Appl.* **28,** 54–58.

van Drongelen, W., and van Loon, J. A. (1980). Inheritance of gustatory sensitivity in $F_1$ progeny of crosses between *Yponomeuta cagnagellus* and *Y. malinellus* (Lepidoptera). *Entomol. Exp. Appl.* **28,** 199–203.

Waldbauer, G. P., and Fraenkel, G. (1961). Feeding on normally rejected plants by maxillectomized larvae of the tobacco hornworm, *Protoparce sexta* (Lepidoptera, Sphingidae). *Ann. Entomol. Soc. Am.* **54,** 477–485.

Wasserman, S. S., and Futuyma, D. J. (1981). Evolution of host plant utilization in laboratory populations of the southern cowpea weevil, *Callosobruchus maculatus* Fabricius (Coleoptera: Bruchidae). *Evolution* **35,** 605–617.

# Index

## B



## C

## D

## E



## F

## G



## H

## Q



## R



## S

## T

## U



## V



## W



## X



## Y

## Z